Advanced Glasses and Glass-Ceramics

Advanced Glasses and Glass-Ceramics

Editors

Georgiy Shakhgildyan
Michael I. Ojovan

Basel • Beijing • Wuhan • Barcelona • Belgrade • Novi Sad • Cluj • Manchester

Editors
Georgiy Shakhgildyan
Department of Glass and
Glass-Ceramics
Mendeleev University of
Chemical Technology
Moscow
Russia

Michael I. Ojovan
Department of
Radiochemistry
Lomonosov Moscow State
University
Moscow
Russia

Editorial Office
MDPI
St. Alban-Anlage 66
4052 Basel, Switzerland

This is a reprint of articles from the Special Issue published online in the open access journal *Ceramics* (ISSN 2571-6131) (available at: www.mdpi.com/journal/ceramics/special_issues/Y15S3F09W2).

For citation purposes, cite each article independently as indicated on the article page online and as indicated below:

Lastname, A.A.; Lastname, B.B. Article Title. *Journal Name* **Year**, *Volume Number*, Page Range.

ISBN 978-3-7258-0528-0 (Hbk)
ISBN 978-3-7258-0527-3 (PDF)
doi.org/10.3390/books978-3-7258-0527-3

Cover image courtesy of Natalia Klimenko

© 2024 by the authors. Articles in this book are Open Access and distributed under the Creative Commons Attribution (CC BY) license. The book as a whole is distributed by MDPI under the terms and conditions of the Creative Commons Attribution-NonCommercial-NoDerivs (CC BY-NC-ND) license.

Contents

About the Editors vii

Preface ix

Valeri V. Poluektov, Vladislav A. Petrov, Michael I. Ojovan and Sergey V. Yudintsev
Uranium Retention in Silica-Rich Natural Glasses: Nuclear Waste Disposal Aspect
Reprinted from: *Ceramics* **2023**, *6*, 69, doi:10.3390/ceramics6020069 1

Alexander Grebenchukov, Olga Boytsova, Alexey Shakhmin, Artem Tatarenko, Olga Makarevich, Ilya Roslyakov, et al.
Infrared and Terahertz Spectra of Sn-Doped Vanadium Dioxide Films
Reprinted from: *Ceramics* **2023**, *6*, 79, doi:10.3390/ceramics6020079 13

Saule Dyussembekova, Ekaterina Trusova, Sergey Kichanov, Kiril Podbolotov and Denis Kozlenko
A Study of PbF_2 Nanoparticles Crystallization Mechanism in Mixed Oxyde-Fluoride Glasses
Reprinted from: *Ceramics* **2023**, *6*, 93, doi:10.3390/ceramics6030093 24

Alyona Vepreva, Dmitry Dubovtsev, Daria Krainova, Yulia Chetvertnykh, Semyon Belyakov, Nailya Saetova and Anton Kuzmin
Barium Silicate Glasses and Glass–Ceramic Seals for YSZ-Based Electrochemical Devices
Reprinted from: *Ceramics* **2023**, *6*, 81, doi:10.3390/ceramics6030081 33

Olga Dymshits, Anastasia Bachina, Irina Alekseeva, Valery Golubkov, Marina Tsenter, Svetlana Zapalova, et al.
Phase Transformations upon Formation of Transparent Lithium Alumosilicate Glass-Ceramics Nucleated by Yttrium Niobates
Reprinted from: *Ceramics* **2023**, *6*, 92, doi:10.3390/ceramics6030092 49

Andrey S. Naumov, Georgiy Yu. Shakhgildyan, Nikita V. Golubev, Alexey S. Lipatiev, Sergey S. Fedotov, Roman O. Alekseev, et al.
Tuning the Coefficient of Thermal Expansion of Transparent Lithium Aluminosilicate Glass-Ceramics by a Two-Stage Heat Treatment
Reprinted from: *Ceramics* **2024**, *7*, 1, doi:10.3390/ceramics7010001 67

Israel R. Montoya Matos
Influence of Alkali Metal Ions on the Structural and Spectroscopic Properties of Sm^{3+}-Doped Silicate Glasses
Reprinted from: *Ceramics* **2023**, *6*, 109, doi:10.3390/ceramics6030109 81

Ksenia Serkina, Irina Stepanova, Aleksandr Pynenkov, Maria Uslamina, Konstantin Nishchev, Kirill Boldyrev, et al.
Bismuth-Germanate Glasses: Synthesis, Structure, Luminescence, and Crystallization
Reprinted from: *Ceramics* **2023**, *6*, 97, doi:10.3390/ceramics6030097 92

Leonid Yu. Mironov, Dmitriy V. Marasanov, Mariia D. Sannikova, Ksenia S. Zyryanova, Artem A. Slobozhaninov and Ilya E. Kolesnikov
Formation and Photophysical Properties of Silver Clusters in Bulk of Photo-Thermo-Refractive Glass
Reprinted from: *Ceramics* **2023**, *6*, 96, doi:10.3390/ceramics6030096 106

Dmitry Butenkov, Anna Bakaeva, Kristina Runina, Igor Krol, Maria Uslamina, Aleksandr Pynenkov, et al.
New Glasses in the $PbCl_2$–PbO–B_2O_3 System: Structure and Optical Properties
Reprinted from: *Ceramics* **2023**, *6*, 83, doi:10.3390/ceramics6030083 . . . . **119**

Sonja Smiljanić, Uroš Hribar, Matjaž Spreitzer and Jakob König
Water-Glass-Assisted Foaming in Foamed Glass Production
Reprinted from: *Ceramics* **2023**, *6*, 101, doi:10.3390/ceramics6030101 . . . . **136**

Lydia V. Ermakova, Valentina G. Smyslova, Valery V. Dubov, Daria E. Kuznetsova, Maria S. Malozovskaya, Rasim R. Saifutyarov, et al.
Effect of a Phosphorus Additive on Luminescent and Scintillation Properties of Ceramics GYAGG:Ce
Reprinted from: *Ceramics* **2023**, *6*, 91, doi:10.3390/ceramics6030091 . . . . **145**

About the Editors

Georgiy Shakhgildyan

Dr. Georgiy Shakhgildyan, an esteemed Associate Professor at Mendeleev University of Chemical Technology, served as the Guest Editor of the "Advanced Glasses and Glass-Ceramics" Special Issue. Holding a Ph.D. in Chemistry from the same university, his research ambitiously targets the development of innovative glass and glass-ceramic materials for cutting-edge applications in optoelectronics and photonics. A prolific contributor to the scientific community, Dr. Shakhgildyan has authored over 50 publications in prestigious journals and secured multiple patents for his groundbreaking work in glass composition and processing techniques. Recognized as a vanguard in his field, he has been honored with several state research grants and the coveted Moscow government award for young scientists. Driven by a deep commitment to propelling glass science toward a sustainable future, Dr. Shakhgildyan is dedicated to fostering interdisciplinary collaborations that push the boundaries of material science. His work not only advances our understanding of glass and glass-ceramics, but also lays the foundation for future innovations in technology and sustainability.

Michael I. Ojovan

Michael I. Ojovan has been a Professor at Imperial College London, Nuclear Engineer at the International Atomic Energy Agency (IAEA), Associate Reader at the University of Sheffield, UK, Leading Scientist at Lomonosov Moscow State University, and Leading Scientist at the Institute of Geology of Ore Deposits, Petrography, Mineralogy and Geochemistry (IGEM) of the Russian Academy of Sciences. He is the Chief Editor of the journal *Science and Technology of Nuclear Installations*. Ojovan has published 15 monographs, including the "Handbook of Advanced Radioactive Waste Conditioning Technologies" and the three editions of "An Introduction to Nuclear Waste Immobilisation" by Elsevier. He has founded and led the IAEA International Predisposal Network (IPN) and the IAEA International Project on Irradiated Graphite Processing (GRAPA). He is known for the connectivity-percolation theory (CPT) of glass transition, the Douglas-Doremus-Ojovan (DDO) model of viscosity of glasses and melts, theoretical bases of condensed Rydberg matter (RM), metallic and glass-composite materials (GCM) for nuclear waste immobilisation, and self-sinking capsules for investigating the Earth's deep interior.

Preface

With a rich history spanning over 5000 years, glass continues to play a pivotal role in propelling science and technology forward, finding its place in a myriad of human endeavor aspects. This enduring material, characterized by its versatility and capacity for innovation, stands at the forefront of cutting-edge applications, bridging disciplines such as materials science, chemistry and physics. The unique ability of glass to incorporate nearly all chemical elements has fostered an environment of continual discovery, with each year bringing forth new compositions and processing techniques. These advancements pave the way for novel materials that support essential and sustainable technologies. The Special Issue aims to spotlight the expansive and dynamic nature of glass science and technology, offering a platform to explore the material's evolving roles in fostering sustainable human development. This edition focuses on cutting-edge research into the structure and properties of glass and glass-based materials, alongside their innovative applications in critical fields such as photonics, energy and sustainable development. The papers selected for inclusion in this reprint reflect significant progress in glass science, showcasing the material's diverse and groundbreaking uses across various domains. Together, these works not only push the boundaries of our understanding of glass materials and their capabilities, but also underscore the importance of cross-disciplinary collaboration in tackling complex issues, from nuclear waste management to advances in optical technology and the pursuit of sustainability. This compilation marks a significant contribution to the body of literature on glass science and technology, encompassing both the study of materials and the synthesis methods employed. It is a testament to the collective effort and dedication of the international glass community. We extend our heartfelt gratitude to all contributors for their pivotal role in bringing this Special Issue to fruition. As we turn the pages of this reprint, we are reminded of the rich tapestry of knowledge and innovation that defines our field. We are inspired by the promise it holds for the future, confident that the insights shared here will inspire further exploration and discovery. In closing, we express our sincere appreciation of the entire glass community for their collaborative spirit and contributions. This reprint stands as a beacon of collective achievement and a source of inspiration for ongoing and future endeavors in the fascinating world of glass science and technology.

Georgiy Shakhgildyan and Michael I. Ojovan
Editors

Article

Uranium Retention in Silica-Rich Natural Glasses: Nuclear Waste Disposal Aspect

Valeri V. Poluektov, Vladislav A. Petrov, Michael I. Ojovan * and Sergey V. Yudintsev

Institute of Geology of Ore Deposits, Petrography, Mineralogy, and Geochemistry Russian Academy of Sciences, Staromonetny Lane, 35, 119017 Moscow, Russia; vapol@igem.ru (V.V.P.); vlad243@igem.ru (V.A.P.); syud@igem.ru (S.V.Y.)
* Correspondence: m.i.ojovan@gmail.com

Abstract: Uranium-containing glass samples with an age of 140–145 million years were collected within the volcanic rocks of the largest volcanic-related uranium ore deposit in the world. Main features of their composition are high concentrations of silica and uranium, the largest for the rocks of this type. In contrast to this, the ages of fresh (unaltered) low-silica natural glasses of a basic composition (basalts) usually do not exceed a few million years. The volcanic low-silica glass is unstable at longer times and in older ancient rocks is transformed into a crystalline mass. The geochemistry of uranium including the behavior in solids and solutions is similar to that of long-lived transuranic actinides such as radioactive Np and Pu from high-level radioactive waste. This allows uranium to be used as a simulant of these long-lived hazardous radionuclides both at the synthesis and for the study of various nuclear wasteforms: glasses, glass crystalline materials and crystalline ceramics. The data obtained on long-term behavior of natural glasses are of importance for prognosis and validation of stability of nuclear wasteforms disposed of in geological disposal facilities (GDF).

Keywords: volcanic glasses; nuclear waste; uranium retention

Citation: Poluektov, V.V.; Petrov, V.A.; Ojovan, M.I.; Yudintsev, S.V. Uranium Retention in Silica-Rich Natural Glasses: Nuclear Waste Disposal Aspect. *Ceramics* **2023**, *6*, 1152–1163. https://doi.org/10.3390/ceramics6020069

Academic Editor: Anna Lukowiak

Received: 6 April 2023
Revised: 11 May 2023
Accepted: 17 May 2023
Published: 18 May 2023

Copyright: © 2023 by the authors. Licensee MDPI, Basel, Switzerland. This article is an open access article distributed under the terms and conditions of the Creative Commons Attribution (CC BY) license (https://creativecommons.org/licenses/by/4.0/).

1. Introduction

Glasses are solids quenched from liquids without phase separation, which can be avoided using enough rapid cooling rates. For example, metallic glasses are produced using fast cooling whereas silicate glasses are formed on the natural cooling of melts because crystallization in such systems proceeds at very low cooling rates. Upon heating, glasses continuously change their properties to those of a liquid-like state (melt) in contrast to crystals, where such changes occur abruptly at the melting point, T_m. The ranges of solid-like and liquid-like behavior of amorphous materials are divided by glass transition temperature, T_g. Whether a material behaves either as a liquid or a solid depends on the connectivity between its elementary building blocks—atoms, molecules, or clusters. Solids are characterized by a high degree of connectivity whereas structural blocks in melts have a lower connectivity. There is a threshold connectivity determining the T_g in each actual system being a function of composition and logarithmically dependent on cooling rate [1,2]. Being solid-state solutions, glasses are highly tolerant to compositional changes. Properties of glasses change continuously with variation of composition. The high chemical durability and tolerance of glasses to compositional variations as well as the ease of their production by cooling of molten mixtures of substances have determined vitrification to be an effective method of nuclear waste immobilization. Indeed, nuclear waste vitrification provided a high degree of retention of radionuclides and significantly contributed to the increased safety of the storage, transportation and final disposal of high-level nuclear waste, which is a dangerous by-product of the peaceful use of nuclear energy [3–5]. Several countries have operated vitrification facilities for decades, significantly reducing the hazard of environmental contamination arising from highly radioactive liquid waste generated at spent nuclear fuel reprocessing (Table 1).

Table 1. Data on industrial vitrification of high-level nuclear waste [6].

Country, Facilities	Performance Data
France, R7/T7, AVM	8252 tonnes, 291·10^6 TBq to 2019
USA, DWPF, WVDP, WTP	7870 tonnes, 2.7·10^6 TBq to 2012
Russia, EP-500	6200 tonnes, 23.8·10^6 TBq to 2010
UK, WVP	2200 tonnes, 33·10^6 TBq to 2012
Belgium, Pamela	500 tonnes, 0.5·10^6 TBq. Completed.
Japan, Tokai	70 tonnes, 14.8·10^3 TBq to 2007
Germany, Karlsruhe	55 tonnes, 0.8·10^6 TBq. Completed.
India, WIP (1), AVS, WIP (2)	28 tonnes, 9.62·10^3 TBq to 2012
Slovakia, Bohunice	1.53 m^3 to 2012

Silicate glasses have been generically selected as the most reliable wasteform to immobilize high-level nuclear waste (HLW) apart from Russia, which uses phosphate glass and partially the joint Belgium–Germany vitrification program Pamela (Table 2).

Table 2. Composition of HLW glasses, wt% [7].

Country	Plant	Glass Composition
Belgium	Pamela	70.7P_2O_5·7.1Al_2O_3·22.2Fe_2O_3 and 52.7SiO_2·13.2B_2O_3·2.7Al_2O_3·4.6CaO·2.2MgO·5.9Na_2O·18.7 Misc. [1]
France	AVM	46.6SiO_2·14.2B_2O_3·5.0Al_2O_3·2.9Fe_2O_3·4.1CaO·10.0Na_2O·17.2 Misc.
France	R7/T7	54.9SiO_2·16.9B_2O_3·5.9Al_2O_3·4.9CaO·11.9Na_2O·5.5 Mis.
Germany	Karlsruhe	60.0SiO_2·17.6B_2O_3·3.1Al_2O_3·5.3CaO·7.1Na_2O·6.9 Mis.
India	WIP	30.0SiO_2·20.0B_2O_3·25.0PbO·5.0Na_2O·20.0 Mis.
India	AVS	34.1SiO_2·6.4B_2O_3·6.2TiO_2·0.2Na_2O·9.3MnO·43.8 Mis.
Japan	Tokai	46.7SiO_2·14.3B_2O_3·5.0Al_2O_3·3.0CaO·9.6Na_2O·21.4 Mis.
Russia	EP500	53.3P_2O_5·15.8Al_2O_3·1.6Fe_2O_3·23.5Na_2O·5.8 Misc.
UK	WVP	47.2SiO_2·16.9B_2O_3·4.8Al_2O_3·5.3MgO·8.4Na_2O·17.4 Misc.
US	DWPF	49.8SiO_2·8.0B_2O_3·4.0Al_2O_3·1.0CaO·1.4MgO·8.7Na_2O·27.1 Misc.
US	WVDP	45.8SiO_2·8.4B_2O_3·6.1Al_2O_3·11.4Fe_2O_3·1.4MgO·9.1Na_2O·17.8 Mis.
US	WTP	50.0SiO_2·20.0B_2O_3·5.0Al_2O_3·25.0Na_2O

[1] Miscellaneous, including oxides of radioactive waste.

Nuclear waste management is a mature internationally regulated industry that deals with all aspects of nuclear waste generated as dangerous by-products of the application of nuclear energy. Although there are not unresolved problems with controlling and handling nuclear waste, many scientific areas related to nuclear waste management focus the attention of experts. These activities include the analysis of long-term behavior of vitrified nuclear waste in deep geological facilities (GDF) after its disposal [8]. Natural glass-like materials can therefore be useful as analogues of vitrified nuclear waste in extrapolating short-term experiments to longer time frames and projections about the long-term safety of disposal [9,10]. Natural glasses to a certain degree similar in composition to that of the HLW glasses (Table 2) are found in nature and have been subjected to conditions similar to that expected if the GDF is becoming flooded with water and the canisters are breached, allowing groundwater to react with the glassy wasteform. The study of natural glasses with ages of millions of years can provide the necessary link between theoretical models on the long-term vitreous nuclear waste stability, study of historical vitrified material with ages of up to 1800 years [11,12] and laboratory tests aiming to provide information on the long-term durability of a glassy wasteform [13].

Volcanic rocks such as those with a silica-rich (rhyolites, obsidians) or low silica content (basalts) composition are usually used as natural analogues for borosilicate (B-Si) vitreous nuclear wasteforms to predict their alteration in a long-time perspective [9,14]. Studies are devoted to the analysis of their alteration from the mineralogical point of view and mechanical stability. Investigations were also performed on the geochemistry of U, Th and rare earth elements (REE) in the rocks, but these studies are mainly connected with the problem of the source of uranium of deposits [15–17]; for example, for uranium ores of Transbaikalia, Russia [18,19]. The common oxidation states of uranium in nature are +4 and +6, whilst in synthetic ceramics and glasses, three valent forms of uranium are observed: +4, +5 and +6, which are also typical for Np and Pu [20,21]. Therefore, uranium can serve as a good radioactive simulant for the investigation of behavior of transuranic actinides both at the synthesis and aging of nuclear wasteforms.

It is generally accepted that vitreous basic volcanic rocks (basalts) can be used as natural analogues (counterparts) of borosilicate glass [22–24] due to the similarity in mechanisms of their alterations. Both basaltic and borosilicate glasses are characterized by a decrease in leaching rate with time by three to five orders of magnitude in comparison with the initial rate. An extremely low rate of long-term dissolution for both kinds of glasses is controlled by the very slow diffusion through the altered layer. Basaltic glasses have low concentrations of uranium, usually at the order of a few ppm (10^{-4} wt%), thus the analysis of its behavior is rather difficult. For this reason, only some silica-rich (acid) volcanic rocks with uranium content up to 140 ppm [25,26] are of special interest for characterization of uranium behavior in natural glasses as analogues of synthetic vitreous materials, including wasteforms for actinide-containing waste. The description of such uranium-rich glasses is given below.

2. Natural Radioactive Glasses

The rocks investigated were formed from melts that naturally and rapidly cooled enough to form glasses rather than forming crystalline rocks. The high chemical resistance of silicate glasses allows them to remain stable and almost unaltered in the environment for many millions of years. For example, the highly siliceous volcanic glass found in the Novogodnee deposit (the U-glass) with an age of 135–145 million years (My) contains uranium at concentrations ranging from tens to 140 ppm [18,19,25,26] that is an order of magnitude superior to the average concentration of this element in such rocks.

The Novogodnee deposit is situated within a reducing geochemical environment of the volcano-sedimentary cover of a volcano caldera in the southern part of the Streltsovska uranium ore field (Figure 1).

Figure 1. Novogodnee deposit (shown by the red star) is located in the central part of the Streltsovka ore field within the Streltsovka caldera (within the yellow contour) with three paleovolcanoes: 1—South-Western, 2—Krasniy Kamen and 3—Maliy Tulukuiy (see details in [25,26]).

The volcanic caldera of Jurassic–Cretaceous age (145–140 Ma) has a diameter of about 20 km and a total area of 180 km^2 comprising 20 uranium deposits. The host rocks are up to 1.4 km thick of volcano-sedimentary accumulation within the caldera lying on a granitic Proterozoic–Paleozoic basement. Ore concentrates of U (thousands ppm) are observed within veins, as sub-vertical stockworks and along stratiform layers in the volcanic and sedimentary units. The mineral phases of U present are oxide (pitchblende), silicates (coffinite) and titanate (brannerite). The total resources are about 280 ktU with average uranium content in ore equal to 0.2 wt%.

The uranium deposits can serve as natural analogues of currently deployed deep geological disposal facilities (GDF) in stable geological conditions [27–31]. Indeed, features, events and processes (FEPs), which occurred during previous periods of time of such deposits, can be analyzed so that by reversing the time, one can appraise the most probable scenario of GDF evolution [32]. A sheetlike body of highly preserved rhyolite–rhyodacite glass of an obsidian–perlite type rich in U was found in the Novogodnee deposit at a depth of 300 m from the surface of the Earth. Data on the behavior of glasses with uranium at the Novogodnee deposit are essential to investigations of actinide migration paths, mobilization, retardation, redistribution and accumulation under existing redox conditions.

The unaltered volcanic high-silica glass in this field was found to contain abnormally high concentrations of uranium detected by f-radiography (XRF). Uranium retardation can be caused by sorption. This factor, together with the deformational alteration and partial devitrification of volcanic glass, is of particular significance at the Novogodnee deposit [26]. The local reducing barriers are formed along the periphery of a sheetlike body of obsidian–perlites near mineralized and open fractures and in intensely altered cataclastic zones. The results of detailed micro-studies indicate an intensive redistribution of uranium with the accumulation of its ultra-high contents up to 500–1000 ppm. Such local favorable conditions within the sheetlike body are formed in the presence of sorption-intensive mineral phases (leucoxene–hematite aggregate) and a higher degree of fracturing of glasses. Here, the ability of the sorption-intensive phases to retard uranium was fully manifested.

3. Bulk Characterization of the Natural Uranium Volcanic Glass

The sheetlike body of rhyolite–rhyodacite volcanic glass has a zonal structure with unaltered massive and fluidal obsidian–perlites (U-glasses) in the core, surrounded by the zone with volcanic bombs and rock fragments developed in the central part (Figure 2 and Tables 3 and 4).

Figure 2. *Cont.*

Figure 2. Samples and position of volcanic glass bed-like body in felsite rhyolite layer: (**a**) the top of the fresh obsidian–perlite volcanic glasses bed; (**b**) well-preserved obsidian–perlite glasses; (**c**) bottom of the cataclastic and altered glasses bed; and (**d**) sample of U-glass NY5 from depth of 300 m.

Chemical analyses of the rock samples were performed in the Centre for Collective Use of Institute of Geology of Ore Deposits, Petrography, Mineralogy, and Geochemistry Russian Academy of Sciences—Analitika using an X-ray fluorescence analysis and Axios mAX&PAN analytical spectrometer for rock-forming elements as well as ICP mass spectrometry, Nexion 2000 c for rare elements (Tables 3 and 4).

Table 3. Compositions of glasses (rock-forming elements, wt%), description in the text.

No.	SiO_2	TiO_2	Al_2O_3	ΣFeO	MnO	MgO	CaO	Na_2O	K_2O	P_2O_5	S	F	LOI [1]
NY1	70.77	0.12	11.51	1.26	0.081	0.18	0.75	4.16	4.24	0.01	0.03	0.32	6.04
NY5-1	71.38	0.13	11.23	1.29	0.085	0.06	0.44	5.83	2.55	<0.01	<0.01	-	6.59
NY5-2	71.78	0.17	11.41	1.37	0.041	0.29	1.80	2.58	5.24	0.02	<0.01	-	5.05
NY5-3	71.76	0.12	11.30	1.24	0.071	0.16	0.84	4.33	3.66	<0.01	0.03	0.07	6.22
NY22-1	71.50	0.12	11.20	1.23	0.085	0.11	0.49	5.66	2.49	<0.01	<0.01	0.05	6.63
NY22-2	71.51	0.12	11.32	1.26	0.072	0.19	0.95	3.96	3.78	<0.01	<0.01	0.05	6.44
NY19-2	66.81	0.13	0.13	1.30	0.062	0.24	2.42	2.81	3.26	0.02	0.06	0.29	10.48
NY15	61.60	0.14	0.14	2.43	0.194	3.93	2.27	1.32	1.92	0.02	0.14	0.53	12.05
NY17	62.76	0.15	0.15	1.89	0.536	1.24	2.65	1.05	2.72	0.01	0.06	0.25	12.64
NY18	64.57	0.14	0.14	2.29	0.226	1.37	2.31	1.41	1.99	0.02	0.06	0.33	12.02
NY12	59.59	0.14	0.14	1.53	0.207	4.03	2.55	0.55	2.30	0.01	0.09	0.83	14.90
NY7-2	66.80	0.14	0.14	1.42	0.259	1.01	2.15	1.58	3.78	0.02	0.03	0.45	9.41

[1] LOI—loss on ignition. Dash (-): the element was not determined.

Table 4. Compositions of glasses (description in the text) for rare and radioactive elements [1], ppm.

No.	Li	Rb	Sr	Cs	Co	Zr	Nb	Mo	Ba	Ta	Pb	Th	U
NY1	96.9	808.9	173.4	933.6	2.5	197.1	55.0	5.2	23.1	4.7	42.6	45.9	25.8
NY5-1	89.9	752.5	33.2	291.9	0.3	189.8	51.6	7.9	6.5	4.5	28.9	44.6	19.1
NY5-2	245.4	410.5	343.1	758.3	0.9	195.1	56.6	6.6	71.2	4.4	35.8	56.7	23.5
NY5-3	132.6	790.5	155.2	753.4	0.4	199.7	55.3	7.9	18.9	4.7	38.5	53.6	23.7
NY22-1	94.4	913.1	56.8	605.3	0.2	196.8	49.1	6.5	5.3	2.7	30.3	32.4	17.8
NY22-2	127.2	666.3	149.3	827.7	0.3	198.7	51.7	5.7	10.8	3.9	28.2	15.9	12.8
NY19-2	63.7	179.2	680.6	1201.7	17.1	183.6	48.9	3.8	36.2	12.3	23.3	36.3	18.4
NY15	77.4	152.7	493.9	1485.2	402.1	205.3	47.5	2.5	237.5	4.6	57.0	52.1	9.7
NY17	134.2	186.9	3244.2	744.6	3.9	228.5	62.2	-	137.3	5.9	81.1	70.7	15.6
NY18	87.6	182.2	595.8	1715.8	59.4	212.5	118.1	-	186.4	1032.4	17.8	65.8	14.2
NY12	538.7	140.9	443.5	273.2	69.7	211.3	59.0	2.9	102.3	5.5	49.3	53.5	15.3
NY7-2	172.9	238.8	1057.8	966.3	4.8	195.6	53.2	0.6	61.4	4.5	45.4	45.3	15.2

[1] Dash (-): the element was not determined.

Volcanic rock samples NY1, NY5 and NY22 are seen as glasses being relatively fresh and unaltered whereas samples NY19-2, NY15, NY17 and NY18 are characterized as intensively altered and devitrified glasses. Samples NY18 and NY19-2 are cataclastic and altered glasses. The fresh obsidian–perlites NY5 (such as shown below in Figure 2) and NY22 were divided into three groups: NY5-1 and NY22-1 (obsidian), NY5-2 and NY22-2 (perlite) and NY5-3 (interbedding of obsidian and perlite).

Devitrification of the glass led to the formation of crystallites within glasses such as hair trichites, globulites and aggregates of scopulites. At terminal stages of crystallization, spherulites and microlites were formed (Figure 3, see also [26]).

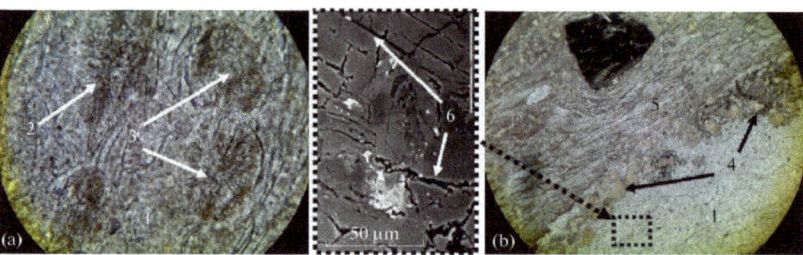

Figure 3. Massive (1) and fluidal (5) glasses of Novogodnee deposit with successive devitrification stages: hair trichites, (**a**)—globulites; (3)—scopulite aggregates; (2)—spherulites; and (4)—crystallites. The long side of thin section is 1.09 mm (**a**) or 2.78 mm (**b**), parallel nicols; The BSE image is shown in the center of the figure.

Volcanic glasses of the Novogodnee deposit, although being significantly older than a hundred million years, still contain uranium at relatively high concentrations, retaining it both in the glassy and crystalline phases. The distribution of uranium in volcanic glasses was studied with f-radiography using polished thin sections covered by film of special track detectors, which were irradiated by thermal neutrons [26]. The f-radiography method is based on the fission of uranium nuclei in a nuclear reactor under irradiation by neutrons and allows, with high sensitivity and accuracy (mass sensitivity threshold of the order of $\cdot 10^{-10}$) detecting the distribution of uranium, as well as determining local and total concentrations. The fundamental possibility of determining the uranium content is based on the fact that fragments of the spontaneous or forced fission of nuclei of heavy elements leave destruction areas (tracks) in the environment, which can be detected under a microscope after the chemical etching of the surface. Qualitative assessments of uranium distribution and calculation of its contents were performed by analyzing micro-images from lavsan detectors, following the methods and software developed at the Institute of Geology of Ore Deposits, Petrography, Mineralogy, and Geochemistry Russian Academy of Sciences. Polished thin sections of the samples mounted on quartz glasses and covered by lavsan track detectors were irradiated at the National Research Nuclear University (MEPhI) Atomic Center with a thermal neutrons fluence of 3×10^{16} neutrons/cm^2 for low (1–3 ppm) uranium contents, and 3×10^{14} and 4×10^{13} neutrons/cm^2 for high (>3–5 ppm) and very high (>500–1000 ppm) contents, respectively. Relatively high content of uranium is hence characteristic for both unaltered and altered volcanic rocks of the Novogodnee deposit including partly devitrified glasses (Table 5).

Table 5. Distribution of uranium in volcanic glasses.

Samples	Number of Sites Studied	Content of U, ppm		Variation Coefficient
		Average	Range	
Relatively unaltered massive and fluidal glasses [1]				
Fresh glass (NY22-1)	8	25.26	23.97–27.47	5.31
Fresh glass (NY5-1)	9	19.30	18.17–21.03	4.97
Initial devitrified I (NY22-1)	9	17.85	17.30–19.11	3.59
Initial devitrified II (NY5-1)	9	14.12	12.54–15.18	6.94
Altered and devitrified glasses				
Altered I glass (NY23-1)	9	14.75	13.12–17.75	10.78
Altered II glass (NY0-1)	9	11.42	9.49–12.19	8.06
Altered III glass (NY2-1)	9	5.34	5.06–6.62	10.86
Altered IV glass (NY26-1)	9	1.72	1.60–2.69	26.16
Area with HEM (NY26-1) [2]	5	39.55	32.12–55.33	22.66

[1] The term fluidal glass denotes a glass where the direction of melt flow that occurred before its vitrification is revealed from the image of glass. [2] Areas uniformly impregnated with sorption-capacious hematite (HEM) were identified in brown volcanic glass (perlite) with elevated uranium content detected.

4. Discussion

Two aspects of the geochemical and mineralogical studies of natural glasses are of particular interest. The first one is related to their study as a potential source of useful elements, e.g., U, in the uranium ore deposits in volcanic-related mineral systems. The second direction concerns the use of volcanic rocks as analogues of vitreous actinide wasteforms used for HLW immobilization in assessing their long-term stability under underground disposal conditions. Data revealed through such type analyses can be a significant addition to existing short-term testing protocols used in laboratories [33].

Silicate glasses have been used for some five decades to immobilize HLW radionuclides including residual uranium and plutonium isotopes remnant from spent nuclear fuel reprocessing [3–6,12,34]. They were also proposed for the immobilization of nuclear fuel containing lava, which resulted from the Chernobyl catastrophe [35]. HLW glasses used in practice are characterized by a high content of crystalline phases within their vitreous body [3,5,36]; moreover, there is a technological trend to increase the content of crystalline phases [37], accounting for advantages offered by glass-crystalline wasteforms [38–40]. Volcanic glasses investigated do also contain crystalline phases, which could be either initially formed at magmatic melt (volcanic lava) solidification or be products of glass devitrification. The most important structural feature of them would then be the speciation of uranium within glass and the crystalline phase's structure. Although uranium can exist in the oxidation states U^{6+}, U^{5+} and U^{4+} in alkali borosilicate glasses, most of the uranium (~90%) occurs as U^{6+} with a small amount of U^{5+} under standard melting conditions using the air atmosphere. Moreover, U^{4+} is not observed, with the exception of strongly reducing conditions. The U^{6+} occurs in the form of the uranyl species UO_2^{2+} with an additional four or five equatorial oxygens coordinating to the uranium cation [41–44]. It is supposed that uranium acts essentially as a glass intermediate within the alkali silicate glass following the rearrangements [44]:

$$UO_3 + 4Si(O^-)\left(O_{\frac{1}{2}}\right)_3 + 4M^+ \rightarrow U^{6+}(=O)_2\left(O_{\frac{1}{2}}\right)_6 + 4Si\left(O_{\frac{1}{2}}\right)_4 + 4M^+, \quad (1)$$

where M^+ is the alkali ion. Each mole of UO_3 thus requires two moles of alkali for charge balance, which results in little structural rearrangement of the glass where uranium occupies sites in the interstices of the glass network related to the alkali channels [44]. This allows for the high solubility of UO_3 in silicate glasses, which is in a borosilicate glass above 40 wt% under an atmosphere of air, whereas under reducing conditions it is only about 10 wt% [4,41,42]. The authors of [43] have shown that two distinct first neighbor distances occur for the U–O correlations, the first being at 1.8 ± 0.05 Å having 1.9 ± 0.2 oxygens and the second at 2.2 ± 0.05 Å with 3.7 ± 0.2 oxygen atoms. They have observed significant second neighbor atomic pair correlations between uranium and the network formers (Si, B) and the modifiers such as Na, concluding that uranium ions take part in the network forming, and that this may be the reason for the observed good stability of uranium containing silicate glasses and their high hydrolytic resistance. It is supposed that uranium atoms in the borosilicate glass are connected through an oxygen atom with the network former and modifier atoms, forming a network structure that consists of mixed tetrahedral made of SiO_4, units, tetrahedral made of BO_4 units and BO_3 trigonal units, which are partly connected by uranium atoms [43,44]. Obtained data and the similarity in the U^{6+} environments in layered alkali uranates and in silicate glasses have allowed the authors of [44] to suggest the possible structural model for silicate glasses containing uranium where the uranyl ions sit in the alkali channels of the modified random network of silicate glasses (see e.g., [2]).

Radiation from decaying radionuclides gradually alters the glass structure, leading to changes in its properties including valence states and migration potential of radionuclides [7,45,46]. Although the content of uranium in the above-described glasses is considered anomalously high for natural volcanic samples, their radioactivity is much lower than that of high-level waste [3–5] when radiation-induced effects are evident [7,47]. The direct determination of uranium valent states in volcanic glasses thus remains a key challenge and its complexity is partly caused by both its content, which is about tens of ppm, and complex composition that are complicating the characterization. This will assist in resolving the unique and scientifically most difficult aspect of nuclear waste management, aiming to extrapolate wasteform short-term laboratory testing results to the long-time periods ranging from thousands to millions of years [48].

Migration of uranium is initiated by glass corrosion in contact with groundwater. Silicate glasses corrode in water via two main processes—(a) diffusion-controlled ion exchange and (b) hydrolysis [5]. The ion exchange mechanism involves the mutual diffusion and exchange of a cation in glass with a proton (probably in the form of H_3O^+) from water via the reaction:

$$(\equiv Si\text{-}O\text{-}M)_{glass} + H_2O \leftrightarrow (\equiv Si\text{-}O\text{-}H)_{glass} + M\text{-}OH, \qquad (2)$$

In dilute near-neutral solutions, the ion exchange controls the initial release of cations and at relatively low temperatures and not very high pH it can dominate over hydrolysis for many hundreds or even thousands of years [49]. Ion exchange reactions cause selective leaching of cations and are characterized by the normalized leaching rate given by:

$$NR_{xi} = \rho(D_i/\pi t)^{1/2} = \rho 10^{-pH/2}[\kappa D_{0H}/C_i(0)\, \pi t]^{1/2} \exp(-E_{di}/2RT), \qquad (3)$$

where ρ (g/cm^3) is the glass density, D_i (cm^2/day) is the effective interdiffusion coefficient, E_{di} (J/mol) is the interdiffusion activation energy (e.g., British magnox waste glass has E_{di} = 36 kJ/mol [49]), R is the universal gas constant R = 8.314 J/mol, T (K) is the temperature, D_{0H} (cm^2/day) is the pre-exponential coefficient in the diffusion coefficient of protons (hydronium ions) in glass, $C_i(0)$ (mol/L) is the concentration of cations at the surface of the glass and κ is a constant that relates the concentration of protons in glass to the concentration of protons in water, i.e., to the pH of the solution. The leaching rate decreases with time t, as $t^{-1/2}$, and the lower the pH of the contacting solution, the higher the $NR_{x,i}$. Hydrolysis, which is a near-surface reaction of hydroxyl ions with the glass network, destroys it, and this leads to congruent dissolution of all glass components into contacting water and subsequent deposition of silica gel layers as secondary products on the surface of glass. Hydrolysis occurs through the reaction [5]:

$$(\equiv Si\text{-}O\text{-}Si\equiv)_{glass} + H_2O \leftrightarrow 2(\equiv Si\text{-}O\text{-}H). \qquad (4)$$

This leads to the complete dissolution of the glass network (disordered glass lattice) and the formation of ortho-silicic acid, H_4SiO_4. Glass hydrolysis is characterized by a normalized dissolution rate [5]:

$$NR_{xi} = \rho r_c = \rho k a^{-\eta}[1 - (Q/K)^\sigma]\exp(-E_a/RT), \quad (5)$$

where r_c is the normalized rate of dissolution measured in unit cm/day, k is the characteristic rate constant, a is the activity of hydrogen ions (protons), η is the pH-dependent exponent (typically $\eta \sim 0.5$), E_a is the activation energy of hydrolysis (e.g., British magnox waste glass has $E_a = 60$ kJ/mol [49]), Q is the product of the ionic activity of the rate control reaction, K is the equilibrium constant of this reaction and σ is the order of the reaction. The rate of hydrolytic dissolution does not depend on time. The higher the pH of the contacting water, the higher the NR_H.

The total corrosion rate NR_i (normalized with respect to the content of component labelled (i)) is determined by the sum of the contributions from (a) ion exchange, (b) hydrolysis and (c) very rapid (so-called instantaneous) dissolution of radionuclides from the glass surface, i.e., surface contaminants [5,49]:

$$NR_i = NR_{X,i} + NR_H + N\Phi_{si}, \quad (6)$$

where the factor $N\Phi_{si}$ accounts for instantaneous dissolution. When the pH changes, the ion exchange rate ($NR_{X,i}$) changes as $10^{-pH/2}$, while the hydrolysis rate (NR_H) is proportional to $10^{pH/2}$, so the dependence of the glass corrosion rate (NR_i) on pH appears as a U-shaped curve with a minimum for neutral solutions [50]. In saturated solutions, Q → K, which is typical for glasses in a geological environment such as volcanic glasses in a volcano caldera or vitrified HLW in a GDF; therefore $NR_H \to 0$. That is, in this case, the corrosion of the glass is determined by diffusion-controlled ion exchange. In dilute aqueous solutions, ion exchange is characteristic for the initial stages of corrosion, and hydrolysis for the later stages. The time it takes to transition from one mechanism to another depends on the composition of the glass and environmental conditions such as pH and temperature. The higher the temperature, the shorter the transition time [49]. The newly formed glass surfaces, due to the cracking of large glass blocks, will follow this pattern of transition of corrosion mechanisms from the initial selective corrosion via ion exchange to the late mechanism controlled by hydrolysis; thus, the overall composition of components leached out of glasses is complex. The analysis shows that the typical rate of corrosion via the hydrolysis (r_c) of borosilicate glasses used to immobilize nuclear waste is about 0.1 μm/y [51]. Without contact of glass with water, the radionuclides including uranium will remain retained within the glass body structure.

5. Conclusions

Volcanic-type uranium deposits of the Streltsovskoye ore field provide a unique opportunity to study uranium-bearing volcanic glasses under various redox conditions. Joint geological–structural, mineralogical–geochemical, petrophysical, hydrogeochemical and isotope–geochemical monitoring studies of fracture veins and atmospheric waters have been conducted for more than 20 years and continue at present. It is shown that the rocks of the Novogodnee deposit are unique objects, which can be used for studying the conditions, migration paths, migration mechanisms and accumulation of uranium in different structural settings under varying redox conditions. Despite devitrification processes within the geological time frames, the ancient volcanic rocks within the large Streltsovka uranium ore field were found to contain large-size blocks of natural glasses having the highest content of silica. The silica-rich rocks located in the southern part of the ore field, formed about 140–145 million years ago, have retained uranium for geological time scales, preventing its migration out of its body. Silica-rich volcanic glasses still confine the uranium as a proof of the high reliability of vitreous and glass crystalline wasteforms, based on silicate glasses, used as nuclear waste immobilizing matrices at geological time scales.

Author Contributions: Conceptualization, V.A.P. and S.V.Y.; methodology, V.V.P.; formal analysis, V.V.P. and M.I.O.; investigation, V.V.P.; resources, V.A.P. and S.V.Y.; data curation, V.V.P.; writing—original draft preparation, M.I.O.; writing—review and editing, V.A.P., V.V.P. and S.V.Y.; supervision, V.A.P.; project administration, V.A.P. and S.V.Y.; funding acquisition, V.A.P. All authors have read and agreed to the published version of the manuscript.

Funding: The research was performed under an assignment task for IGEM RAS supported by the Ministry of Science and Higher Education of the Russian Federation.

Institutional Review Board Statement: Not applicable.

Informed Consent Statement: Not applicable.

Data Availability Statement: Data supporting reported results can be found within the paper and references provided.

Acknowledgments: Chemical analyses of the rock samples were performed in the Centre for Collective Use of IGEM RAS (IGEM—Analitika).

Conflicts of Interest: The authors declare no conflict of interest.

References

1. Ojovan, M.I. Glass formation. In *Encyclopedia of Glass Science, Technology, History, and Culture*; Richet, P., Conradt, R., Takada, A., Dyon, J., Eds.; Wiley: Hoboken, NJ, USA, 2021; Volume I, Chapter 3.1; pp. 249–259, 1568p.
2. Ojovan, M.I. The Modified Random Network (MRN) Model within the Configuron Percolation Theory (CPT) of Glass Transition. *Ceramics* **2021**, *4*, 121–134. [CrossRef]
3. Jantzen, C.M. Development of glass matrices for HLW radioactive wastes. In *Handbook of Advanced Radioactive Waste Conditioning Technologies*; Ojovan, M., Ed.; Woodhead: Cambridge, UK, 2011; pp. 230–292.
4. Gin, S.; Jollivet, P.; Tribet, M.; Peuget, S.; Schuller, S. Radionuclides containment in nuclear glasses: An overview. *Radiochim. Acta* **2017**, *105*, 927–959. [CrossRef]
5. Ojovan, M.I.; Lee, W.E.; Kalmykov, S.N. *An Introduction to Nuclear Waste Immobilisation*, 3rd ed.; Elsevier: Amsterdam, The Netherlands, 2019; 497p.
6. Ojovan, M.I. Glass is key for nuclear waste immobilisation. *Glass Int.* **2020**, *43*, 77–80.
7. Malkovsky, V.I.; Yudintsev, S.V.; Ojovan, M.I.; Petrov, V.A. The Influence of Radiation on Confinement Properties of Nuclear Waste Glasses. *Sci. Technol. Nucl. Install.* **2020**, *2020*, 8875723. [CrossRef]
8. IAEA. *Scientific and Technical Basis for Geological Disposal of Radioactive Wastes*; Technical Reports Series 413; IAEA: Vienna, Austria, 2003.
9. Ewing, R.C.; Roed, G. Natural analogues: Their application to the prediction of the long-term behavior of nuclear waste glasses. *Mat. Res. Soc. Symp. Proc.* **1987**, *84*, 67–83. [CrossRef]
10. Laverov, N.P.; Omel'yanenko, B.I.; Yudintsev, S.V.; Stefanovsky, S.V. Confinement matrices for low- and intermediate-level radioactive waste. *Geol. Ore Depos.* **2012**, *54*, 1–16. [CrossRef]
11. Verney-Carron, A.; Gin, S.; Librourel, G.A. Fractured roman glass block altered for 1800 years in seawater: Analogy with nuclear waste glass in a deep geological repository. *Geochim. Cosmochim. Acta* **2008**, *72*, 5372–5385. [CrossRef]
12. Nava-Farias, L.; Neeway, J.J.; Schweiger, M.J.; Marcial, J.; Canfield, N.L.; Pearce, C.I.; Peeler, D.K.; Vicenzi, E.P.; Kosson, D.S.; Delapp, R.C.; et al. Applying laboratory methods for durability assessment of vitrified material to archaeological samples. *Npj Mater. Degrad.* **2021**, *5*, 57. [CrossRef]
13. McKenzie, W.F. *Natural Glass Analogues to Alteration of Nuclear Waste Glass: A Review and Recommendations for Further Study*; Report UCID—21871 (DE90 013513); Earth Sciences Department, Lawrence Livermore National Laboratory, University of California: Livermore, CA, USA, 1990; 36p.
14. Morgenstein, M.E.; Shettel, D.L. Volcanic Glass as a Natural Analog for Borosilicate Waste Glass. *Mat. Res. Soc. Symp. Proc.* **1994**, *333*, 605–615. [CrossRef]
15. Rosholt, J.N.; Noble, D.C. Mobility of Uranium and Thorium in Glassy and Crystallized Silicic Volcanic Rocks. *Econ. Geol.* **1971**, *66*, 1061–1069. [CrossRef]
16. Zielinski, R.A. Tuffaceous sediments as source rocks for uranium—A case study of the White River Formation, Wyoming. *J. Geochem. Explor.* **1983**, *18*, 285–306. [CrossRef]
17. Cuney, M. Felsic magmatism and uranium deposits. *Bull. Soc. Géol. Fr.* **2014**, *185*, 75–92. [CrossRef]
18. Shatkov, G.A. Streltsovsky type of uranium deposits. *Reg. Geol. Metallog.* **2015**, *63*, 85–96.
19. Andreeva, O.V.; Petrov, V.A.; Poluektov, V.V. Mesozoic Acid Magmatites of Southeastern Transbaikalia: Petrogeochemistry and Relationship with Metasomatism and Ore Formation. *Geol. Ore Depos.* **2020**, *62*, 69–96. [CrossRef]

20. Runde, W. The chemical interactions of actinides in the environment. *Los Alamos Sci.* **2000**, *26*, 338–357.
21. Ewing, R.C. Actinides and radiation effects: Impact on the back-end of the nuclear fuel cycle. *Miner. Mag.* **2011**, *75*, 2359–2377. [CrossRef]
22. Advocat, T.; Jollivet, P.; Crovisier, J.L.; del Nero, M. Long-Term Alteration Mechanisms in Water for SON 68 Radioactive Bo-rosilicate Glass. *J. Nucl. Mater.* **2001**, *298*, 55–62. [CrossRef]
23. Crovisier, J.-L.; Advocat, T.; Dussossoy, J.-L. Nature and role of natural alteration gels formed on the surface of ancient volcanic glasses (Natural analogs of waste containment glasses). *J. Nucl. Mater.* **2003**, *321*, 91–109. [CrossRef]
24. Grambow, B. Nuclear Waste Glasses—How Durable? *Elements* **2006**, *2*, 357–364. [CrossRef]
25. Petrov, V.A. The nature of U behaviour in the processes of transformation of volcanic glasses of different composition. In *Joint ICTP-IAEA International School on Nuclear Waste Vitrification*; Abdus Salam International Centre for Theoretical Physics (ICTP): Trieste, Italy, 2019.
26. Poluektov, V.V.; Petrov, V.A.; Andreeva, O.V. Migration and Sorption of Uranium in Various Redox Conditions on the Example of Volcanic-Related Deposits in the Streltsovka Caldera, SE Transbaikalia. *Geol. Ore Depos.* **2021**, *63*, 29–61. [CrossRef]
27. Alexander, W.; McKinley, I. A review of the application of natural analogues in performance assessment: Improving models of radionuclide transport in groundwaters. *J. Geochem. Explor.* **1992**, *46*, 83–115. [CrossRef]
28. Bruno, J.; Duro, L.; Grivé, M. The applicability and limitations of thermodynamic geochemical models to simulate trace element behaviour in natural waters. Lessons learned from natural analogue studies. *Chem. Geol.* **2002**, *190*, 371–393. [CrossRef]
29. Chapman, N.A.; McKinley, I.G.; Smellie, J.A.T. The Potential of Natural Analogues in Assessing Systems for Deep Disposal of High-Level Radioactive Waste. NAGRA Technical Report Series NTB; NAGRA: Wettingen, Switzerland, 1984; pp. 41–84.
30. Haveman, S.A.; Pedersen, K. Microbially mediated redox processes in natural analogues for radioactive waste. *J. Contamin. Hydrol.* **2002**, *55*, 161–174. [CrossRef]
31. Smellie, J.A.T.; Karlsson, F.; Alexander, W.R. Natural analogue studies: Present status and performance assessment implications. *J. Contamin. Hydrol.* **1997**, *26*, 3–17. [CrossRef]
32. Chapman, N.; Hooper, A. The disposal of radioactive wastes underground. *Proc. Geol. Assoc.* **2012**, *123*, 46–63. [CrossRef]
33. Thorpe, C.L.; Neeway, J.J.; Pearce, C.I.; Hand, R.J.; Fisher, A.J.; Walling, S.A.; Hyatt, N.C.; Kruger, A.A.; Schweiger, M.; Kosson, D.S.; et al. Forty years of durability assessment of nuclear waste glass by standard methods. *Npj Mater. Degrad.* **2021**, *5*, 61. [CrossRef]
34. Bitay, E.; Kacsó, I.; Veress, E. Chemical Durability of Uranium Oxide Containing Glasses. *Acta Mater. Transilv.* **2018**, *1*, 12–18. [CrossRef]
35. Olkhovyk, Y.A.; Ojovan, M.I. Corrosion resistance of Chernobyl NPP lava fuel-containing masses. *Innov. Corros. Mater. Sci.* **2015**, *5*, 36–42. [CrossRef]
36. Rose, P.B.; Woodward, D.I.; Ojovan, M.I.; Hyatt, N.C.; Lee, W.E. Crystallisation of a simulated borosilicate high-level waste glass produced on a full-scale vitrification line. *J. Non-Cryst. Solids* **2011**, *357*, 2989–3001. [CrossRef]
37. Gribble, N.R.; Short, R.; Turner, E.; Riley, A.D. The Impact of Increased Waste Loading on Vitrified HLW Quality and Durability. *Mat. Res. Soc. Symp. Proc.* **2009**, *1193*, 283. [CrossRef]
38. Ojovan, M.I.; Lee, W.E. Glassy and glass composite nuclear wasteforms. *Ceram. Trans.* **2011**, *227*, 203–216.
39. Ojovan, M.I.; Petrov, V.A.; Yudintsev, S.V. Glass Crystalline Materials as Advanced Nuclear Wasteforms. *Sustainability* **2021**, *13*, 4117. [CrossRef]
40. Ojovan, M.I.; Yudintsev, S.V. Glass, ceramic, and glass-crystalline matrices for HLW immobilisation. *Open Ceram.* **2023**, *14*, 100355. [CrossRef]
41. Farges, F.; Ponader, C.W.; Calas, G.; Brown, G.E. Structural environments of incompatible elements in silicate glass/melt systems: II. UIV, UV, and UVI. *Geochim. Cosmochim. Acta* **1992**, *56*, 4205–4220. [CrossRef]
42. Schreiber, H.D.; Balazs, G.B. The chemistry of uranium in borosilicate glasses. Part1. Simple base compositions relevant to the immobilisation of nuclear waste. *Phys. Chem. Glas.* **1982**, *23*, 139.
43. Fábián, M.; Sváb, E.; Zimmermann, M. Structure study of new uranium loaded borosilicate glasses. *J. Non-Cryst. Solids* **2013**, *380*, 71–77. [CrossRef]
44. Connelly, A.; Hyatt, N.; Travis, K.; Hand, R.; Stennett, M.; Gandy, A.; Brown, A.; Apperley, D. The effect of uranium oxide additions on the structure of alkali borosilicate glasses. *J. Non-Cryst. Solids* **2013**, *378*, 282–289. [CrossRef]
45. Ojovan, M.I.; Lee, W.E. Alkali ion exchange in γ-irradiated glasses. *J. Nucl. Mater.* **2004**, *335*, 425–432. [CrossRef]
46. Patel, K.B.; Boizot, B.; Facq, S.P.; Peuget, S.; Schuller, S.; Farnan, I. Impacts of composition and beta irradiation on phase separation in multiphase amorphous calcium borosilicates. *J. Non-Cryst. Solids* **2017**, *473*, 1–16. [CrossRef]
47. Ojovan, M.I. Challenges in the Long-Term Behaviour of Highly Radioactive Materials. *Sustainability* **2022**, *14*, 2445. [CrossRef]
48. Ewing, R.C. Ageing Studies of Nuclear Waste Forms. In *Ageing Studies and Lifetime Extension of Materials*; Mallinson, L.G., Ed.; Springer: Boston, MA, USA, 2001. [CrossRef]
49. Ojovan, M.I.; Pankov, A.; Lee, W.E. The ion exchange phase in corrosion of nuclear waste glasses. *J. Nucl. Mater.* **2006**, *358*, 57–68. [CrossRef]

50. Ojovan, M.I.; Lee, W.E. About U-shaped Glass Corrosion Rate/pH Curves for Vitreous Nuclear Wasteforms. *Innov. Corros. Mater. Sci.* **2017**, *7*, 30–37. [CrossRef]
51. Ojovan, M.I.; Hand, R.J.; Ojovan, N.V.; Lee, W.E. Corrosion of alkali–borosilicate waste glass K-26 in non-saturated conditions. *J. Nucl. Mater.* **2005**, *340*, 12–24. [CrossRef]

Disclaimer/Publisher's Note: The statements, opinions and data contained in all publications are solely those of the individual author(s) and contributor(s) and not of MDPI and/or the editor(s). MDPI and/or the editor(s) disclaim responsibility for any injury to people or property resulting from any ideas, methods, instructions or products referred to in the content.

Article

Infrared and Terahertz Spectra of Sn-Doped Vanadium Dioxide Films

Alexander Grebenchukov [1], Olga Boytsova [2,3], Alexey Shakhmin [1], Artem Tatarenko [3], Olga Makarevich [2], Ilya Roslyakov [3,4], Grigory Kropotov [1] and Mikhail Khodzitsky [1,*]

1. Tydex LLC, 194292 Saint Petersburg, Russia
2. Department of Chemistry, Lomonosov Moscow State University, 119991 Moscow, Russia
3. Department of Materials Science, Lomonosov Moscow State University, 119991 Moscow, Russia
4. Kurnakov Insitute of General and Inorganic Chemistry RAS, 119991 Moscow, Russia
* Correspondence: khodzitskiy@yandex.ru

Abstract: This work reports the effect of tin (Sn) doping on the infrared (IR) and terahertz (THz) properties of vanadium dioxide (VO_2) films. The films were grown by hydrothermal synthesis with a post-annealing process and then fully characterized by X-ray diffraction (XRD), Raman spectroscopy, scanning electron microscopy (SEM), and temperature-controlled electrical resistivity as well as IR and THz spectroscopy techniques. Utilizing $(NH_4)_2SnF_6$ as a Sn precursor allows the preparation of homogeneous Sn-doped VO_2 films. Doping of VO_2 films with Sn led to an increase in the thermal hysteresis width while conserving the high modulation depth in the mid-IR regime, which would be beneficial for the applications of VO_2 films in IR memory devices. A further analysis shows that Sn doping of VO_2 films significantly affects the temperature-dependent THz optical properties, in particular leading to the suppression of the temperature-driven THz transmission modulation. These results indicate Sn-doped VO_2 films as a promising material for the development of switchable IR/THz dichroic components.

Keywords: Sn doping; infrared spectroscopy; terahertz transmission; dichroic optical elements

Citation: Grebenchukov A.; Boytsova O.; Shakhmin A.; Tatarenko A.; Makarevich O.; Roslyakov I.; Kropotov G.; Khodzitsky M. Infrared and Terahertz Spectra of Sn-Doped Vanadium Dioxide Films. *Ceramics* 2023, 6, 1291–1301. https://doi.org/10.3390/ceramics6020079

Academic Editor: Georgiy Shakhgildyan

Received: 26 May 2023
Revised: 10 June 2023
Accepted: 13 June 2023
Published: 15 June 2023

Copyright: © 2023 by the authors. Licensee MDPI, Basel, Switzerland. This article is an open access article distributed under the terms and conditions of the Creative Commons Attribution (CC BY) license (https://creativecommons.org/licenses/by/4.0/).

1. Introduction

The mid-infrared (2–20 µm wavelength range) spectral region attracts attention from both scientific and industrial sectors due to the availability of multiple atmospheric windows and its technological potential in thermal imaging [1], free space communications [2], and chemical and biological molecular sensing [3,4] because of the fingerprint vibrational and rotational motions of molecules within this spectral region. Full utilization of mid-infrared radiation's potential still requires active optical components. Phase change materials such as best known as vanadium dioxide (VO_2) can also be useful for the development of mid-infrared photonic applications, especially when combined with resonant plasmonic structures. In recent years, VO_2 has been widely used as the basis of active metamaterials operating in the mid-infrared range [5–11].

VO_2-based devices' functional performance significantly depends on the morphology, preparation methods, and doping of VO_2 films [12]. The most remarkable property of VO_2 is the multi-stimulus-induced [13] reversible phase transition from a dielectric to a metallic state [14]. This metal–insulator transition (MIT) leads to an abrupt variation in its electric, thermal, and optical properties [15]. There are four main criteria defining the performance of the MIT in VO_2: the phase transition amplitude, the phase transition sharpness, the hysteresis width, and the state stability before and after phase transition. Element doping enables tailoring of these key VO_2 performances for application requirements [16].

The temperature of the MIT can be decreased by doping with high valance metal ions (W^{6+}, Mo^{6+}, and Nb^{5+}) [17–19] or increased by doping with low valence atoms (Fe) [20] from its initial value for undoped VO_2 of 68 °C according to the application requirements. Since the phase transition points for cooling and heating processes are incompatible, this results in thermal hysteresis (ΔT_{MIT}). The thermal hysteresis width can be reduced by doping with titanium [21], niobium [22], and tungsten [23] or increased by doping with boron [23]. The phase transition sharpness is defined as the full width at half maximum (FWHM) of the Gaussian fitted differential $d(Tr)/d(T)$ versus temperature curves. Commonly, VO_2 element doping reduces the phase transition sharpness [24,25], except for doping with SiO_2 [26].

It was reported that Sn-doped VO_2 films fabricated by hydrothermal synthesis with $SnCl_4 \cdot 5H_2O$ as the tin precursor possess an enhanced visible light transmittance [27]. W-Sn co-doped VO_2 films exhibit an improved visible transmittance with a reduced MIT temperature [28].

A dichroic optical component can provide the ability to manipulate radiation differently concerning its frequency band [29,30]. Among the dichroic elements demonstrated thus far, conductive thin films such as indium tin oxide [31–33] and La-doped $BaSnO_3$ [34] have been utilized in near-infrared transparent/terahertz functional devices. However, infrared functional devices with a high terahertz transparency still need to be explored.

In this paper, the potential of Sn-doped VO_2 films prepared by hydrothermal synthesis and a post-annealing process in temperature-driven mid-infrared and terahertz optical modulation is determined. To reveal the effect of the VO_2 dopant on the optical properties in the mid-IR spectral range across the MIT, the Sn doping levels were varied. Given the high modulation depth and increased thermal hysteresis width in the mid-IR range, we envision the application of Sn-doped VO_2 films for adaptive infrared camouflage and optical memory-type devices. Moreover, the revealed temperature-dependent modulation suppression in the THz range is helpful for the development of dichroic optical elements.

2. Experimental Details

2.1. Preparation Of Sn-Doped VO_2 Samples

Sn-doped VO_2 films were deposited on 0.5 mm single crystal r-cut sapphires substrates polished on one-side (r-Al_2O_3 Monocrystal Co., Ltd., Stavropol, Russia) by hydrothermal synthesis [35]. Vanadium precursors were synthesized using vanadium pentoxide (V_2O_5) and oxalic acid ($H_2C_2O_4 \cdot 2H_2O$) as starting materials. A mixture of ethylene glycol (EG) and deionized (DI) water was selected as a solvent. Sn-doped vanadium dioxide was obtained by adding hexafluorostannate (($NH_4)_2SnF_6$) as a doping agent.

For producing an aqueous V^{4+}-containing solution, V_2O_5 and $H_2C_2O_4 \cdot 2H_2O$ were mixed in a molar ratio of 1:3 in DI water with continuous magnetic stirring for 6 h at 80 °C. Thereafter, the required amount of EG (DI water/EG = 1:1 V/V) was added. The calculated amount of $(NH_4)_2SnF_6$ was dissolved in a DI/EG solution of V^{4+}. As a result, a precursor solution with different concentrations of tin was obtained. Concentrations of 1% and 1.5% of tin were chosen for the synthesis. This precursor was diluted with the DI/EG solvent to obtain a V^{4+} cation concentration of 3.125 mmol/L.

Sn-doped VO_2 (M_1) films on r-Al_2O_3 substrates were synthesized with hydrothermal deposition with a post-annealing process. Prior to deposition, r-Al_2O_3 crystals (0.55 × 1.5 cm^2) were cleaned with DI water and acetone. Then, the substrates were placed into a high-density 25 mL polyparaphenol (PPL)-lined hydrothermal synthesis autoclave reactor in a vertical position using a Teflon holder. Thereafter, the precursor solution was transferred into the PPL cup with a filling ratio of 0.60 and sealed hermetically in a stainless autoclave. The autoclave was kept at 180 °C for 20 h and then cooled down to room temperature naturally. The films deposited on the substrates were cleaned with DI water and acetone several times and dried for 30 min at room temperature. Post-annealing was performed in an argon gas atmosphere (3 mbar, Ar flow (3.5 L/h)) in two steps. The first step at 400 °C for 30 min was intended to remove any EG residues. On the second annealing, the temperature was increased to 600 °C for 60 min.

Based on the Sn concentration, the samples were denoted as S0 (undoped VO_2), S1 (1% Sn), and S2 (1.5% Sn).

2.2. Characterization Studies

The phase purity and crystallinity of VO_2 films were analyzed by X-ray diffraction (XRD, Rigaku SmartLab) with Cu Kα (λ = 1.54046 Å). The diffraction data were recorded in the 2θ range of 20–80° with a resolution of 0.02° at a speed of 5 °/min. The surface morphology of the films and their thickness were characterized by scanning electron microscopy (SEM) using a Carl Zeiss NVision 40 electron microscope. Raman scattering measurements were performed using a Renishaw InVia spectrometer with a 514 nm 20 mW defocused excitation laser source (20 µm spot) at room temperature. The electrical properties of the films were measured with a standard four-probe method in the temperature range of 25–90 °C using a Keithley 2700 multimeter. The temperature-dependent infrared transmittance in the wavelength range of 1.5–8 µm was investigated using a Bruker Vertex 70 Fourier spectrometer. Finally, the terahertz transmission in the frequency range of 0.1–1 THz was measured using a Menlo Systems TERA K8 terahertz time-domain (THz-TDS) spectroscopy system. All the temperature-dependent optical characterizations were performed with a Peltier-based homemade temperature control system.

3. Results and Discussion

3.1. Structural and Morphological Analysis

Figure 1 shows surface morphology SEM images of Sn-doped VO_2 films with different Sn contents. Doped and undoped VO_2 films exhibited uniform homogeneous coverage of the substrate with quasi-spherical grains. Doping with Sn (Figure 1b,c) led to a significant increase in the quasi-spherical grain size, while various doping levels had a minor effect on film morphology.

Figure 1. SEM morphology view of the VO_2 films: (**a**) S0, (**b**) S1, and (**c**) S2.

The SEM cross-sectional images shown in Figure 2 indicate that VO_2 doping with Sn leads to an increase in film thickness. All the doped films have a thickness lying in the range of 170–200 nm, while the undoped VO_2 film is 95 nm thick.

The crystalline structures of Sn-doped VO_2 films on sapphire substrates were analyzed by XRD measurements at room temperature as shown in Figure 3.

The XRD results show that no additional phase appears in the XRD pattern after Sn doping. All obtained films are polycrystalline or 200 textured. The diffraction peaks are typical of VO_2(M), ICDD PDF#43-1051. This indicates that doping with Sn does not significantly change the lattice constants of VO_2 films. However, the XRD peak with an angular position of 36.9° corresponding to the (200) VO_2 (M1) crystalline plane slightly shifts towards a lower angle with an increase in the Sn dopant.

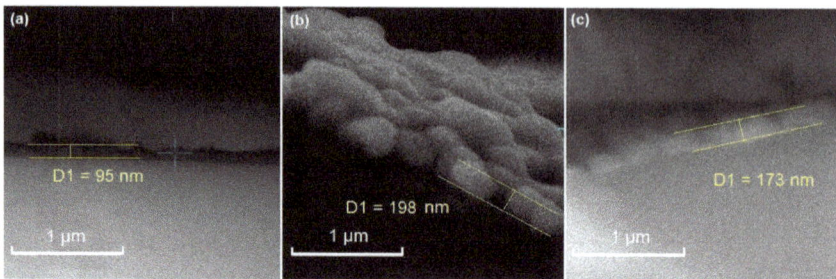

Figure 2. SEM cross-sectional view of the VO$_2$ films: (**a**) S0, (**b**) S1, and (**c**) S2.

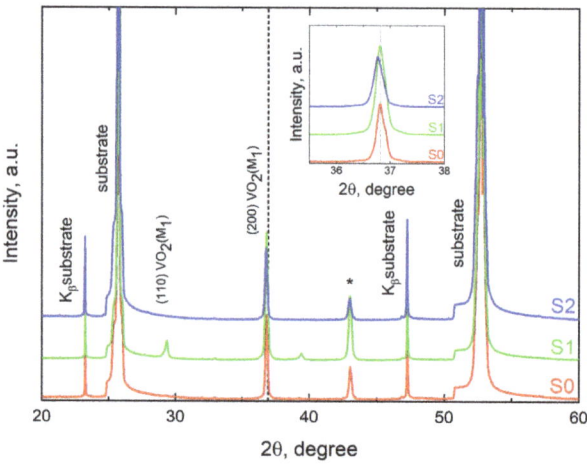

Figure 3. XRD spectra of undoped and Sn-doped VO$_2$ films. The marker "*" indicates the reflection from the stainless sample table.

The typical Raman signature of monoclinic VO$_2$ (M1) was obtained for undoped and Sn-doped VO$_2$ samples (Figure 4).

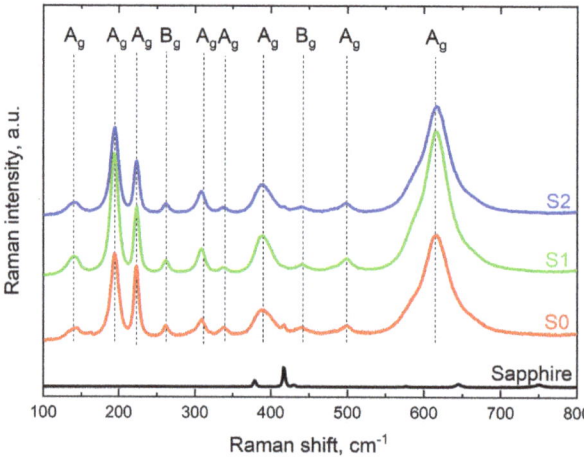

Figure 4. Raman spectra of the sapphire substrate and undoped and Sn-doped VO$_2$ films.

Raman scattering peaks position were identified at 143 (A_g), 195 (A_g), 224 (B_g), 262 (B_g), 309 (A_g), 340 (A_g), 391 (A_g), 442 (B_g), 499 (A_g), and 614 (A_g) cm^{-1}, which clearly conforms with the typical pattern [36].

3.2. Electrical and IR Optical Properties

Figure 5 shows the resistance–temperature hysteresis loops of the undoped and Sn-doped VO$_2$ samples.

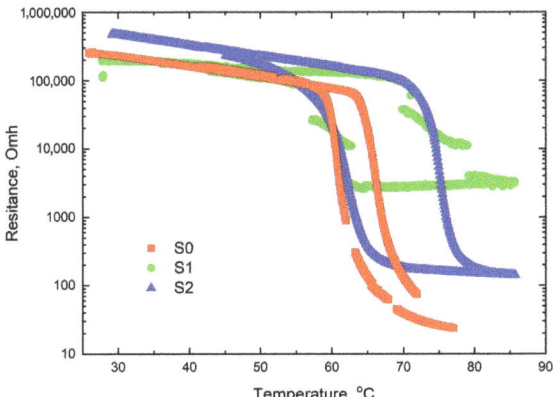

Figure 5. Electrical resistance of the VO$_2$ films as a function of temperature during heating and cooling cycles.

The resistance of the undoped sample dropped by almost 4 orders of magnitude across the phase transition. Sn-doping of VO$_2$ films results in an increase in the overall resistance, MIT temperature growth, and widening of thermal hysteresis loops. Moreover, with increasing Sn content, the magnitude of resistance variation tends to decrease. The incorporation of isovalent Sn^{4+} ions into VO$_2$ does not lead to significant changes in carrier concentrations. However, doping of VO$_2$ generally increases the defect concentration and leads to a more distorted lattice, which as a consequence reduces the phase transition amplitude [12].

Figure 6 represents the infrared transmission of the bare sapphire substrate and VO$_2$ films on sapphire substrates during the heating and cooling processes. It should be noted that the optical properties of the sapphire substrate between 20 °C and 90 °C do not show a significant change as reported in [37].

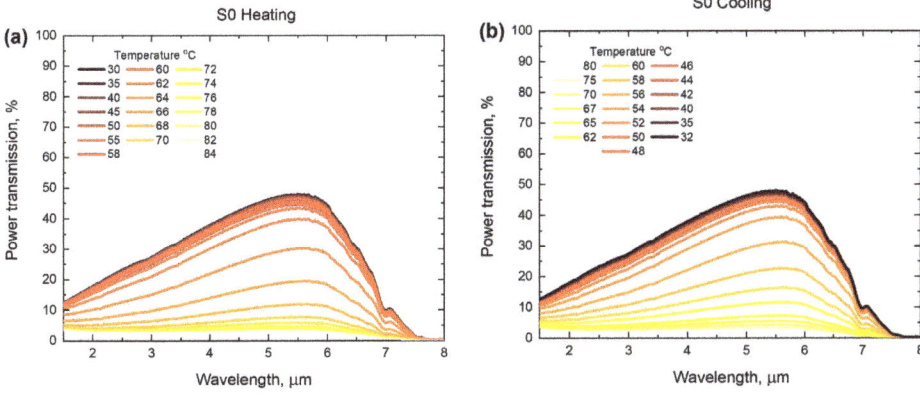

Figure 6. *Cont.*

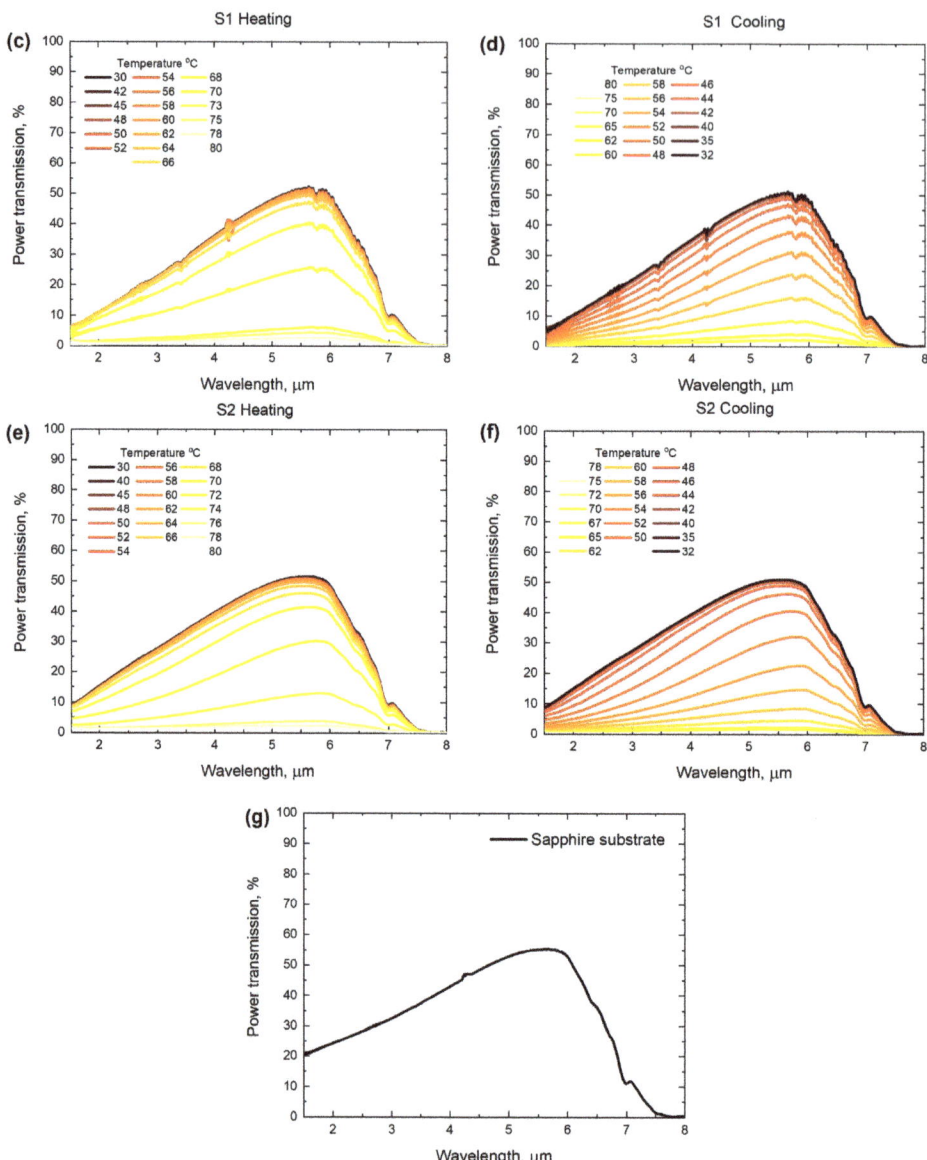

Figure 6. Infrared transmission spectra of the undoped VO_2 film (**a,b**), Sn-doped VO_2 films (**c–f**) on sapphire substrates during heating and cooling cycles, and the bare sapphire substrate (**g**).

The largest transmission variation takes place at 5.6 µm, which coincides with the maximum substrate transmission. For further analysis, the hysteresis loops of IR transmission for undoped and Sn-doped VO_2 films were obtained by collecting the transmittance of films at a fixed wavelength of 5.6 µm as shown in Figure 7. The hysteresis loops of IR transmission through VO_2 films on the sapphire substrate were normalized by transmission through the bare sapphire substrate. In order to quantitatively investigate the IR properties of VO_2 films under a phase transition, the corresponding first-order derivative curves (dTr/dT) of transmission variation were calculated in the insets of Figure 7.

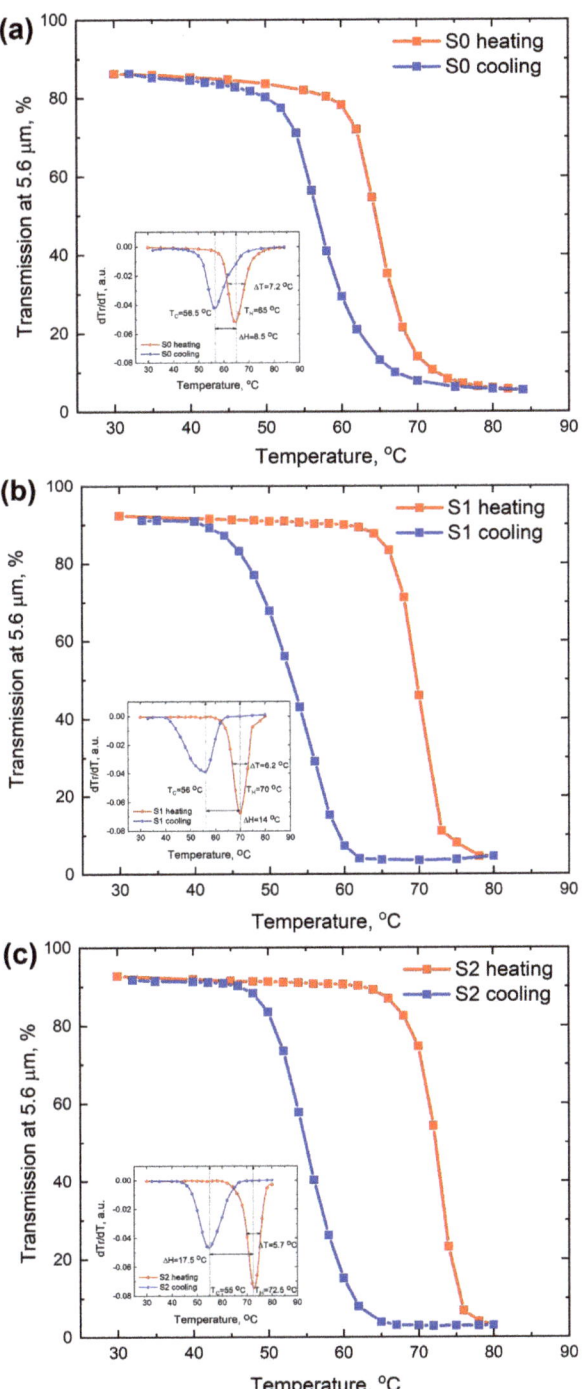

Figure 7. Normalized maximum power transmission at 5.6 μm through undoped (**a**) and Sn-doped (**b**,**c**) VO$_2$ films on a sapphire substrate during heating and cooling cycles.

The temperature-dependent mid-infrared properties of VO$_2$ films are similar to their electrical properties. To gain insight into the phase transition performance of VO$_2$ films with different Sn contents, several criteria were determined. The phase transition temperature was defined as the minima of the differential curves for heating (T_H) and cooling (T_C) processes. The hysteresis width (ΔH) of the phase transition was defined as the difference between phase transition temperatures during heating and cooling processes ($\Delta H = T_H - T_C$). The phase transition sharpness (ΔT) was characterized by the full width at half maximum (FWHM) of the dTr/dT versus the temperature curve. A smaller value of ΔT means a sharper phase transition. The modulation depth was defined as $MD = (T_{cold} - T_{hot})/T_{cold} \times 100\%$, where T_{cold} and T_{hot} are the IR transmission before and after the phase transition, respectively. The detailed parameters of the IR hysteresis loops are summarized in Table 1.

Table 1. Parameters of hysteresis loops at 5.6 µm for VO$_2$ films.

Sample	MD, %	ΔH, °C	ΔT, °C
S0	93.7	8.5	7.2
S1	95	14	6.2
S2	96.8	17.5	5.7

As seen from Table 1, with increasing Sn content, the width of the thermal hysteresis loop (ΔH) is significantly raised from 8.5 °C to 17.5 °C (sample S2). Moreover, the MD is increased from 93.7% to 96.8% and the ΔT is reduced from 7.2 °C to 5.7 °C when the Sn content increases from 0% to 1.5%. Previous reports have indicated that the grain size and grain boundary play important roles in tailoring the thermal hysteresis width [12]. Such a large hysteresis width is preferable for the development of optical-memory-type devices with a stationary memory state [38].

3.3. Thz Optical Properties

The optical transmission of VO$_2$ films with different Sn doping contents in the THz range of 0.1–1 THz was measured at 25 °C and 85 °C, respectively. The corresponding substrate-normalized THz spectra are shown in Figure 8.

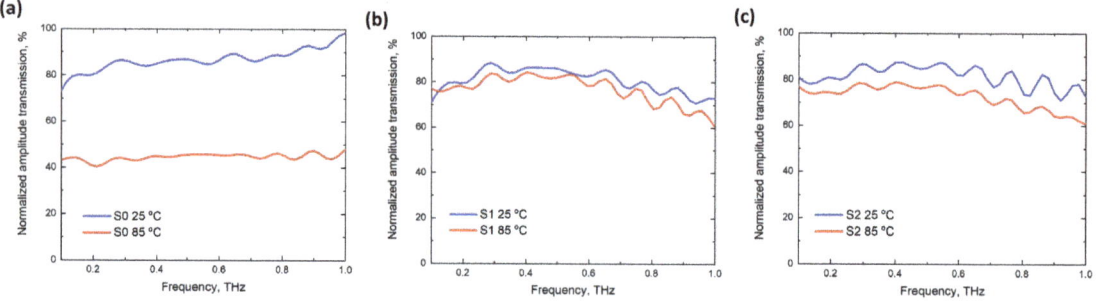

Figure 8. Normalized terahertz transmission spectra through undoped (**a**) and Sn-doped VO$_2$ films on a sapphire substrate (**b**,**c**).

The only undoped sample S0 demonstrates an obvious change in THz transmission between the two states (Figure 8a). With the addition of Sn, the amplitude modulation of THz transmission falls from an average of 49.5% to 2.9% and 10.3% for 1% and 1.5% Sn contents, respectively. An optimal Sn doping level of 1% allows for achieving the largest THz modulation damping. This is consistent with the observed reduction in conductivity after the phase transition for Sn-doped VO$_2$ films as seen from the electrical behavior in Figure 5. A similar relationship between electrical resistance and THz transmission has also been reported in [39,40]. This phenomenon can be attributed to the emergence of barriers between VO$_2$ grains upon dopant insertion. At the same time, the IR optical properties are less sensitive to the interface between the grains. Therefore, the phase transition amplitude for IR transmission varies by a small amount with Sn doping of VO$_2$. The observed reduction in the temperature-driven THz amplitude modulation in conjunction with high IR transmission modulation for Sn-doped VO$_2$ films can be considered as a basis for the development of dichroic optical elements. This feature can be utilized for the separation of generated THz radiation from the initial mid-infrared spectral part in intense THz pulse generation using two-color filamentation techniques [41,42].

4. Conclusions

In summary, a series of VO$_2$ films with different Sn doping contents were prepared on a sapphire substrate by hydrothermal synthesis and a post-annealing process. It was revealed that using (NH$_4$)$_2$SnF$_6$ as a Sn precursor allows producing homogeneous Sn-doped VO$_2$ films. For IR transmission, the hysteresis width of VO$_2$ films can be increased to 17.5 °C by Sn doping. For THz transmission, a suppression of the temperature-driven modulation after Sn doping is observed. This work provides a new mode for the development of dichroic optical components, e.g., a temperature-switchable infrared element with transparency in the THz range.

Author Contributions: Conceptualization, M.K.; methodology, O.B.; resources, G.K.; measurement, A.G., A.S. and I.R.; sample fabrication, A.T. and O.M.; writing—original draft preparation, A.G.; writing—review and editing, A.G., O.B. and M.K.; supervision, M.K.; funding acquisition, G.K. All authors have read and agreed to the published version of the manuscript.

Funding: This research received no external funding. This work was performed using the resources of Tydex LLC.

Data Availability Statement: The data that support the findings of this study are available from the corresponding author upon reasonable request.

Acknowledgments: Experimental studies were carried out on the equipment of the core shared research facilities "Composition, structure and properties of structural and functional materials" of the NRC "Kurchatov Institute"—CRISM "Prometey". SEM was performed using the equipment of the JRC PMR IGIC RAS. The synthesis of films was realized with assistance LLC UGO.

Conflicts of Interest: The authors declare that they have no known competing financial interests or personal relationships that could have appeared to influence the work reported in this paper.

Abbreviations

The following abbreviations are used in this manuscript:

Sn	Tin
VO$_2$	Vanadium dioxide
IR	Infrared
THz	Terahertz
XRD	X-ray diffraction
SEM	Scanning electron microscopy
MIT	Metal–insulator transition
EG	Ethylene glycol
DI	Deionized
THz-TDS	Terahertz time-domain spectroscopy
FWHM	Full width at half maximum

References

1. Tittl, A.; Michel, A.K.U.; Schäferling, M.; Yin, X.; Gholipour, B.; Cui, L.; Wuttig, M.; Taubner, T.; Neubrech, F.; Giessen, H. A switchable mid-infrared plasmonic perfect absorber with multispectral thermal imaging capability. *Adv. Mater.* **2015**, *27*, 4597–4603. [CrossRef]
2. Pang, X.; Ozolins, O.; Jia, S.; Zhang, L.; Schatz, R.; Udalcovs, A.; Bobrovs, V.; Hu, H.; Morioka, T.; Sun, Y.T.; et al. Bridging the Terahertz Gap: Photonics-assisted Free-Space Communications from the Submillimeter-Wave to the Mid-Infrared. *J. Light. Technol.* **2022**, *40*, 3149–3162. [CrossRef]
3. Lin, H.; Luo, Z.; Gu, T.; Kimerling, L.C.; Wada, K.; Agarwal, A.; Hu, J. Mid-infrared integrated photonics on silicon: A perspective. *Nanophotonics* **2018**, *7*, 393–420. [CrossRef]
4. Fang, Y.; Ge, Y.; Wang, C.; Zhang, H. Mid-infrared photonics using 2D materials: Status and challenges. *Laser Photonics Rev.* **2020**, *14*, 1900098. [CrossRef]
5. Shu, F.Z.; Wang, J.N.; Peng, R.W.; Xiong, B.; Fan, R.H.; Gao, Y.J.; Liu, Y.; Qi, D.X.; Wang, M. Electrically Driven Tunable Broadband Polarization States via Active Metasurfaces Based on Joule-Heat-Induced Phase Transition of Vanadium Dioxide. *Laser Photonics Rev.* **2021**, *15*, 2100155. [CrossRef]
6. Erçağlar, V.; Hajian, H.; Özbay, E. VO_2–graphene-integrated hBN-based metasurface for bi-tunable phonon-induced transparency and nearly perfect resonant absorption. *J. Phys. Appl. Phys.* **2021**, *54*, 245101. [CrossRef]
7. Liu, L.; Kang, L.; Mayer, T.S.; Werner, D.H. Hybrid metamaterials for electrically triggered multifunctional control. *Nat. Commun.* **2016**, *7*, 13236. [CrossRef] [PubMed]
8. Dicken, M.J.; Aydin, K.; Pryce, I.M.; Sweatlock, L.A.; Boyd, E.M.; Walavalkar, S.; Ma, J.; Atwater, H.A. Frequency tunable near-infrared metamaterials based on VO_2 phase transition. *Opt. Express* **2009**, *17*, 18330–18339. [CrossRef]
9. Kats, M.A.; Blanchard, R.; Genevet, P.; Yang, Z.; Qazilbash, M.M.; Basov, D.; Ramanathan, S.; Capasso, F. Thermal tuning of mid-infrared plasmonic antenna arrays using a phase change material. *Opt. Lett.* **2013**, *38*, 368–370. [CrossRef]
10. Liu, Z.M.; Li, Y.; Zhang, J.; Huang, Y.Q.; Li, Z.P.; Pei, J.H.; Fang, B.Y.; Wang, X.H.; Xiao, H. Design and fabrication of a tunable infrared metamaterial absorber based on VO_2 films. *J. Phys. Appl. Phys.* **2017**, *50*, 385104. [CrossRef]
11. Kats, M.A.; Sharma, D.; Lin, J.; Genevet, P.; Blanchard, R.; Yang, Z.; Qazilbash, M.M.; Basov, D.; Ramanathan, S.; Capasso, F. Ultra-thin perfect absorber employing a tunable phase change material. *Appl. Phys. Lett.* **2012**, *101*, 221101. [CrossRef]
12. Xue, Y.; Yin, S. Element doping: A marvelous strategy for pioneering the smart applications of VO_2. *Nanoscale* **2022**, *14*, 11054–11097. [CrossRef] [PubMed]
13. Ke, Y.; Wang, S.; Liu, G.; Li, M.; White, T.J.; Long, Y. Vanadium dioxide: The multistimuli responsive material and its applications. *Small* **2018**, *14*, 1802025. [CrossRef]
14. Morin, F. Oxides which show a metal-to-insulator transition at the Neel temperature. *Phys. Rev. Lett.* **1959**, *3*, 34. [CrossRef]
15. Liu, K.; Lee, S.; Yang, S.; Delaire, O.; Wu, J. Recent progresses on physics and applications of vanadium dioxide. *Mater. Today* **2018**, *21*, 875–896. [CrossRef]
16. Shi, R.; Shen, N.; Wang, J.; Wang, W.; Amini, A.; Wang, N.; Cheng, C. Recent advances in fabrication strategies, phase transition modulation, and advanced applications of vanadium dioxide. *Appl. Phys. Rev.* **2019**, *6*, 011312. [CrossRef]
17. Ji, H.; Liu, D.; Cheng, H. Infrared optical modulation characteristics of W-doped VO_2 (M) nanoparticles in the MWIR and LWIR regions. *Mater. Sci. Semicond. Process.* **2020**, *119*, 105141. [CrossRef]
18. Lv, X.; Chai, X.; Lv, L.; Cao, Y.; Zhang, Y.; Song, L. Preparation of porous Mo-doped VO_2 films via atomic layer deposition and post annealing. *Jpn. J. Appl. Phys.* **2021**, *60*, 085501. [CrossRef]
19. Guan, S.; Souquet-Basiège, M.; Toulemonde, O.; Denux, D.; Penin, N.; Gaudon, M.; Rougier, A. Toward room-temperature thermochromism of VO_2 by Nb doping: Magnetic investigations. *Chem. Mater.* **2019**, *31*, 9819–9830. [CrossRef]
20. Victor, J.L.; Gaudon, M.; Salvatori, G.; Toulemonde, O.; Penin, N.; Rougier, A. Doubling of the phase transition temperature of VO_2 by Fe doping. *J. Phys. Chem. Lett.* **2021**, *12*, 7792–7796. [CrossRef]
21. Chen, S.; Liu, J.; Wang, L.; Luo, H.; Gao, Y. Unraveling mechanism on reducing thermal hysteresis width of VO_2 by Ti doping: A joint experimental and theoretical study. *J. Phys. Chem. C* **2014**, *118*, 18938–18944. [CrossRef]
22. Nishikawa, M.; Nakajima, T.; Kumagai, T.; Okutani, T.; Tsuchiya, T. Adjustment of thermal hysteresis in epitaxial VO_2 films by doping metal ions. *J. Ceram. Soc. Jpn.* **2011**, *119*, 577–580. [CrossRef]
23. Yano, A.; Clarke, H.; Sellers, D.G.; Braham, E.J.; Alivio, T.E.; Banerjee, S.; Shamberger, P.J. Toward high-precision control of transformation characteristics in VO_2 through dopant modulation of hysteresis. *J. Phys. Chem. C* **2020**, *124*, 21223–21231. [CrossRef]
24. Gao, Z.; Liu, Z.; Ping, Y.; Ma, Z.; Li, X.; Wei, C.; He, C.; Liu, Y. Low metal–insulator transition temperature of Ni-doped vanadium oxide films. *Ceram. Int.* **2021**, *47*, 28790–28796. [CrossRef]
25. Zou, Z.; Zhang, Z.; Xu, J.; Yu, Z.; Cheng, M.; Xiong, R.; Lu, Z.; Liu, Y.; Shi, J. Thermochromic, threshold switching, and optical properties of Cr-doped VO_2 thin films. *J. Alloys Compd.* **2019**, *806*, 310–315. [CrossRef]
26. Schläefer, J.; Sol, C.; Li, T.; Malarde, D.; Portnoi, M.; Macdonald, T.J.; Laney, S.K.; Powell, M.J.; Top, I.; Parkin, I.P.; et al. Thermochromic VO_2-SiO_2 nanocomposite smart window coatings with narrow phase transition hysteresis and transition gradient width. *Sol. Energy Mater. Sol. Cells* **2019**, *200*, 109944. [CrossRef]
27. Zhao, Z.; Liu, Y.; Wang, D.; Ling, C.; Chang, Q.; Li, J.; Zhao, Y.; Jin, H. Sn dopants improve the visible transmittance of VO_2 films achieving excellent thermochromic performance for smart window. *Sol. Energy Mater. Sol. Cells* **2020**, *209*, 110443. [CrossRef]

28. Zhao, Z.; Liu, Y.; Yu, Z.; Ling, C.; Li, J.; Zhao, Y.; Jin, H. Sn–W Co-doping improves thermochromic performance of VO$_2$ films for smart windows. *ACS Appl. Energy Mater.* **2020**, *3*, 9972–9979. [CrossRef]
29. Naftaly, M. *Terahertz Metrology*; Artech House: Boston, MA, USA 2015.
30. THz Spectral Splitters. Available online: https://www.tydexoptics.com/products/thz_optics/thz_splitter (accessed on 5 June 2023).
31. Bauer, T.; Kolb, J.; Löffler, T.; Mohler, E.; Roskos, H.; Pernisz, U. Indium–tin–oxide-coated glass as dichroic mirror for far-infrared electromagnetic radiation. *J. Appl. Phys.* **2002**, *92*, 2210–2212. [CrossRef]
32. Lai, W.; Yuan, H.; Fang, H.; Zhu, Y.; Wu, H. Ultrathin, highly flexible and optically transparent terahertz polarizer based on transparent conducting oxide. *J. Phys. D Appl. Phys.* **2020**, *53*, 125109. [CrossRef]
33. Chen, C.W.; Lin, Y.C.; Chang, C.H.; Yu, P.; Shieh, J.M.; Pan, C.L. Frequency-dependent complex conductivities and dielectric responses of indium tin oxide thin films from the visible to the far-infrared. *IEEE J. Quantum Electron.* **2010**, *46*, 1746–1754. [CrossRef]
34. Arezoomandan, S.; Prakash, A.; Chanana, A.; Yue, J.; Mao, J.; Blair, S.; Nahata, A.; Jalan, B.; Sensale-Rodriguez, B. THz characterization and demonstration of visible-transparent/terahertz-functional electromagnetic structures in ultra-conductive La-doped BaSnO$_3$ Films. *Sci. Rep.* **2018**, *8*, 3577. [CrossRef]
35. Ivanov, A.V.; Tatarenko, A.Y.; Gorodetsky, A.A.; Makarevich, O.N.; Navarro-Cia, M.; Makarevich, A.M.; Kaul, A.R.; Eliseev, A.A.; Boytsova, O.V. Fabrication of epitaxial W-doped VO$_2$ nanostructured films for terahertz modulation using the solvothermal process. *ACS Appl. Nano Mater.* **2021**, *4*, 10592–10600. [CrossRef]
36. Shvets, P.; Dikaya, O.; Maksimova, K.; Goikhman, A. A review of Raman spectroscopy of vanadium oxides. *J. Raman Spectrosc.* **2019**, *50*, 1226–1244. [CrossRef]
37. Thomas, M.E.; Joseph, R.I.; Tropf, W.J. Infrared transmission properties of sapphire, spinel, yttria, and ALON as a function of temperature and frequency. *Appl. Opt.* **1988**, *27*, 239–245. [CrossRef] [PubMed]
38. Lu, C.; Lu, Q.; Gao, M.; Lin, Y. Dynamic manipulation of THz waves enabled by phase-transition VO$_2$ thin film. *Nanomaterials* **2021**, *11*, 114. [CrossRef]
39. Wu, X.; Wu, Z.; Ji, C.; Zhang, H.; Su, Y.; Huang, Z.; Gou, J.; Wei, X.; Wang, J.; Jiang, Y. THz transmittance and electrical properties tuning across IMT in vanadium dioxide films by Al doping. *ACS Appl. Mater. Interfaces* **2016**, *8*, 11842–11850. [CrossRef]
40. Ji, C.; Wu, Z.; Wu, X.; Feng, H.; Wang, J.; Huang, Z.; Zhou, H.; Yao, W.; Gou, J.; Jiang, Y. Optimization of metal-to-insulator phase transition properties in polycrystalline VO$_2$ films for terahertz modulation applications by doping. *J. Mater. Chem. C* **2018**, *6*, 1722–1730. [CrossRef]
41. Fedorov, V.Y.; Tzortzakis, S. Optimal wavelength for two-color filamentation-induced terahertz sources. *Opt. Express* **2018**, *26*, 31150–31159. [CrossRef]
42. Koulouklidis, A.D.; Gollner, C.; Shumakova, V.; Fedorov, V.Y.; Pugžlys, A.; Baltuška, A.; Tzortzakis, S. Observation of extremely efficient terahertz generation from mid-infrared two-color laser filaments. *Nat. Commun.* **2020**, *11*, 292. [CrossRef]

Disclaimer/Publisher's Note: The statements, opinions and data contained in all publications are solely those of the individual author(s) and contributor(s) and not of MDPI and/or the editor(s). MDPI and/or the editor(s) disclaim responsibility for any injury to people or property resulting from any ideas, methods, instructions or products referred to in the content.

Article

A Study of PbF$_2$ Nanoparticles Crystallization Mechanism in Mixed Oxyde-Fluoride Glasses

Saule Dyussembekova [1,2], Ekaterina Trusova [3,*], Sergey Kichanov [1], Kiril Podbolotov [4] and Denis Kozlenko [1]

1. Frank Laboratory of Neutron Physics, Joint Institute for Nuclear Research, 141980 Dubna, Russia
2. Institute of Nuclear Physics, Ministry of Energy of the Republic of Kazakhstan, Almaty 050032, Kazakhstan
3. Glass and Ceramics Technology Department, Belarusian State Technological University, 220006 Minsk, Belarus
4. State Scientific Institution "The Physical-Technical Institute of the National Academy of Sciences of Belarus", 220141 Minsk, Belarus
* Correspondence: trusova@belstu.by; Tel.: +375-297674337

Abstract: Samples of nanocrystalline PbF$_2$ glass ceramics were obtained by heat-treating SiO$_2$–GeO$_2$–PbO–PbF$_2$–CdF$_2$ glasses. The Ho$_2$O$_3$ and Tm$_2$O$_3$ doping effects on the structural features of PbF$_2$ nanoparticles were studied using small-angle X-ray scattering and X-ray diffraction methods. The enlargements of the average sizes of nanoparticles and the sizes of local areas of density fluctuations have been found to be correlated with an increase in concentrations of Ho$_2$O$_3$ and Tm$_2$O$_3$ in initial glasses. A variation in the concentrations of Ho$_2$O$_3$ and Tm$_2$O$_3$ does not affect the morphology and fractal dimension of the formed PbF$_2$ nanoparticles.

Keywords: oxyfluoride glasses; glass ceramics; small-angle X-ray scattering; nanoparticles

1. Introduction

Fine-tuning the optical properties of glass ceramics with embedded optical nanoparticles, as well as synthesizing novel vitreous composite materials, can extend their applicability and functionality and show novel methods for their technical applications [1,2]. Transparent glass ceramic materials [1–5] are widely used to increase the efficiency of solar cells as near-infrared light sources, in optical glass fibers and sensors and in elements for laser technology [6,7]. First of all, this is due to the formation of non-linear optical properties [2,8,9] in glass-based materials. Their chemical and thermal stability, and ability to finely control the optical properties and structural characteristics of nanoparticles during synthesis, were reported [5,10,11].

Rare-earth (RE) ions are optically active elements that are sources of effective luminescence, and nanoparticles doped with these ions are characterized by high quantum yields, wide possibilities of tuning optical properties, and noticeable suppression of the effect of concentration quenching [12]. Therefore, luminescent glasses and glass ceramics based on rare-earth ions with a stoichiometric or non-stoichiometric composition are a promising replacement for phosphor single crystals [13].

The formation of predominantly amorphous PbF2 nanoparticles and clusters in the glass matrix is observed [5,14]. The nanoparticles form complex fractal-like structures consisting of semi-regular arrangements that resemble concentration bunches inside the glass material. The PbF$_2$ nanoparticles doped with different RE ions embed well in a glass matrix, which is a source of their optical properties, and greatly enhances the emission yields of the corresponding optical materials. The PbF$_2$ nanoparticles are clustered in complex aggregates of sizes 10–30 nm [5]. Recently, it was shown that PbF$_2$ nanoparticles can be doped with RE ions up to a high doping level of 10 wt.% without luminescence yield losses [15]. The optically active nanoparticles could be formed based on the density fluctuations in the glasses, which can definitely affect the structural properties of the PbF$_2$ nanoparticles. Moreover, the glasses and associated glass fibers doped with RE ions have certain advantages, such as a low melting temperature [5] and high ultraviolet resistance. From this point of view, the up-conversion luminescent glass ceramics with PbF_2 nanoparticles are of particular interest [15–18]. These glass materials have potential applications in the fields of emission displays, cathode-ray tubes and solid-state lighters. The phenomenon of up-conversion luminescence is the joint radiation of a multicomponent system through sequential optical transitions between various optically active ions [14–21]. Thus, as an example, one can cite the processes of up-conversion luminescence of PbF_2 nanoparticles doped with Yb^{3+} and Eu^{3+} ions, in which the emission of infrared radiation is observed during optical pumping of Yb^{3+} ions from the energy levels of Eu^{3+} ions [18–20]. During the high-temperature treatment, crystallization of the nanoparticles of PbF_2 from concentration bunches exists [14]. The crystalline nanoparticles are host systems for rare earth elements that provide the conditions for up-conversion luminescence [18]. At the same time, the growth and uniform distribution of these nanoparticles inside the glasses are proposed. The uniform distribution of nanoparticles is the reason for the suppression of the concentration luminescence quenching characteristic of large clusters of nanoparticles [5,14,17]. Both the crystallization and uniform distribution of nanoparticles determine the high intensity of luminescence observed in the glasses. The combination of RE ions as joint activators of optical centers with up-conversion luminescence excitation mechanisms enhances the optical properties of the glass ceramics [15,18].

For transparent glass ceramic materials, the efficiency of up-conversion luminescence depends not only on the type and concentration of optically active RE ions but also on the type and composition of the glass matrix [5,10,14,18], the chemical nature of the formed nanoparticles, and thermal treatment modes [14]. It should be noted that an important problem in the development of glass-nanoceramics is the optimization of the glass composition, which has effects on the spectral–optical properties of the system. Furthermore, the stability of the glass materials is required when introducing fluorides and oxides of RE elements with a molar concentration of a few percent [3,16,17]. Recently, interest in the studies of oxyfluoride and germanium–gallium glasses doped with thulium and holmium ions has grown [19,21,22]. These ions have optical transitions in the infrared region with a high energy transfer efficiency of the Tm^{3+}-Ho^{3+} process during up-conversion luminescence. The presence of holmium ions in glass nanoceramics makes these materials promising for infrared laser sources with a wavelength of 2 μm [21]. Currently, much attention is paid to optimizing the synthesis of such glass ceramics by selecting the optimal ratio of RE ions for the realization of up-conversion luminescence. Previous studies of germanium–gallium glasses with Tm^{3+}/Ho^{3+} ions indicate a maximum efficiency of up-conversion luminescence of 63% at an initial relative concentration of oxides of 70 Tm_2O_3/15 Ho_2O_3 [21]. It is known that the formation of luminescent nanoparticles in silicate composite systems is associated with the chemical processes between oxides of RE elements and components of the glass matrix [5,14]. However, the structural mechanisms of nanoparticle formation in the glass matrixes are studied less.

It is known that the effectiveness of up-conversion luminescence correlates with the nanoparticle structural characteristics, and the growth of nanoparticles depends on the conditions of glass thermal treatment [14]. From the position of the broad peak on the small-angle neutron scattering curves, it is possible to roughly estimate a shift in the average characteristic distance between clusters in the glass ceramics. Those distances increase by factors of 2–3, which can indicate an increase in the spacing between clusters in the glasses [14,18]. Taking into account the practical aspect of developing glass ceramics based on mixed oxide–fluoride glass matrixes as well as the interest in up-conversion luminescent materials, our work is directed to studying the structural aspects of the formation of nanoparticles containing Tm^{3+} and Ho^{3+} ions in mixed oxyfluoride glasses using small-angle X-ray scattering and X-ray diffraction methods.

2. Materials and Methods

The main problem in the development of transparent glass ceramic materials is the optimization of the initial glass composition, which, on the one hand, imparts high spectral luminescence properties, and, on the other hand, the stability of the vitreous state upon the introduction of fluorides and oxides of different RE elements with molar contents of several percent [5,14,17]. A small addition of thulium oxide improves the color properties of the up-conversion luminescence. The Ho^{3+} and Tm^{3+} ions have been chosen because this joint activation may enhance the color characteristics of the luminescence of the glass ceramics due to the diversity of possible mechanisms of excitation of up-conversion luminescence with the participation of pairs of ions Ho^{3+}–Tm^{3+}. The parent glass matrix is the mixed oxyfluoride vitreous system $2.2SiO_2$–$1.3GeO_2$–$6.9PbO$–$7.6PbF_2$–$2CdF_2$–xHo_2O_3–$y\,Tm_2O_3$. The introduction of Ho_2O_3 and Tm_2O_3 oxides in a certain molar range was performed (Table 1). The selected concentration range of Ho_2O_3 corresponds to the maximum efficiency of up-conversion luminescence in similar glasses [21,23]. Structural studies of glasses doped with a variety of optically active nanoparticles are quite complex tasks due to the possible interaction between the nanoparticle elements and the vitreous material. A lead-containing glass matrix has been chosen to provide the light fusibility of the vitreous system. In particular, the introduction of reagents PbO and PbF_2 as glass-formers is a source of components for PbF_2 nanocrystal formation at lower temperatures of ~400 °C and a reduction of the temperature of glass synthesis to 900 °C. CdF_2 is an additional source of fluorine. It should be noted that in the first step, thoroughly mixed components of glass were fritted to reduce the evaporation of fluorine.

Table 1. Concentration of rare-earth oxides. The oxides concentrations are presented in the molar percent.

Sample Label	Ho_2O_3	Tm_2O_3
N1	0.6	–
N2	0.08	0.8
N3	0.4	0.8
N4	0.6	0.8
N5	-	0.8

The synthesis of glass was performed using traditional technology by melting a mixture that was prepared from pure chemical reagents. All reagents were mixed in corresponding proportions and homogenized by milling. The thoroughly mixed charge was placed into the corundum crucibles, which were put into an electric furnace. The synthesis of glasses was performed at a temperature of 950 °C, with exposure at maximum temperature for 30 min. The short period of 30 min is sufficient to complete the homogenization and refining of the glass sample. Glass components are fritted to reduce the evaporation of fluorine. We believe that the fluorine evaporation is less because the studied glass is fusible.

The glasses were annealed in the electric muffle furnace at 300 °C for 3 h. The X-ray diffraction data have confirmed the amorphous nature of the obtained glass materials. The glasses were heat-treated at 350 °C for 30 h + 360 °C for 50 h in order to form the PbF_2 nanoparticles [5,14].

Small-angle X-ray scattering (SAXS) is a useful technique for nanoscale structural characterization of glass materials [24,25]. In SAXS, structural and spatial information is indirectly obtained from the scattering intensity in the spectral domain, known as the reciprocal space [26]. Therefore, characterizing the structure requires solving the inverse problem of finding a plausible structure model that corresponds to the measured scattering intensity. Small-angle X-ray scattering experiments were performed with a Xeuss 3.0 instrument (XENOCS SAS, Grenoble, France). The radiation was generated by a GeniX3D source (Mo-Kα edge, λ = 0.71078 Å). Small-angle X-ray scattering curves were obtained using an Eiger2 detector at different sample-detector distances from 1 to 4 m. The thickness of the glass samples was 1 mm. The SAXS data were corrected for empty container data.

The analysis of small-angle scattering data was performed in the software package SasView (Version 5.0.6) [27].

The crystalline phase of the up-conversion luminescent nanoparticles in the glass matrix was studied using the X-ray diffraction method with the same Xeuss 3.0 device in diffraction mode, with the detector position at a distance of 0.5 m from the sample. We assumed that the crystal phase relates to the cubic phase of β-PbF_2 with space group $Fm\bar{3}m$ [5,15].

3. Results

The SAXS curves of the studied glass materials are shown in Figure 1. The scattered intensity was detected as a function of the momentum transfer modulus $q = (4\pi/\lambda)\sin\theta$, where θ is the scattering angle and λ is the incident X-ray wavelength [25]. The obtained curves for all samples are similar and have a typical shape for disordered glass systems [5,14,24]. There are small changes in the degree of slope of the curves and their shape, which may correspond to a change in the fractal dimensions of scattering objects. Figure 1b shows the Guinier plots $\{ln(I(q)), q^2\}$, which provide the radius of inertia R_g of the scatterers [24,25]. The glass sample behavior exhibits a non-linear trend towards low q, where aggregation is detected by an upward curve, whereas a downward curve will be typical of particle repulsion [24]. It can be seen that, with an increase in the content of thulium oxide Tm_2O_3, there are noticeable changes in the Guinier graphs, which may indicate a clustering or aggregation of smaller particles [23,24]. On the other hand, the model of several scatters will be correct [5,14]. In this model, when the content of oxides increases, an increase in the average size of large particles or aggregates is expected. Therefore, to analyze the SAXS data, we used a two-particle model, which postulates the contribution to SAXS curves from luminescent nanoparticles, most likely PbF_2:Tm-Ho [14], and from fluctuations in density inside the glass matrix [10]. This model has been used previously in studies of other glass systems [5,14,28].

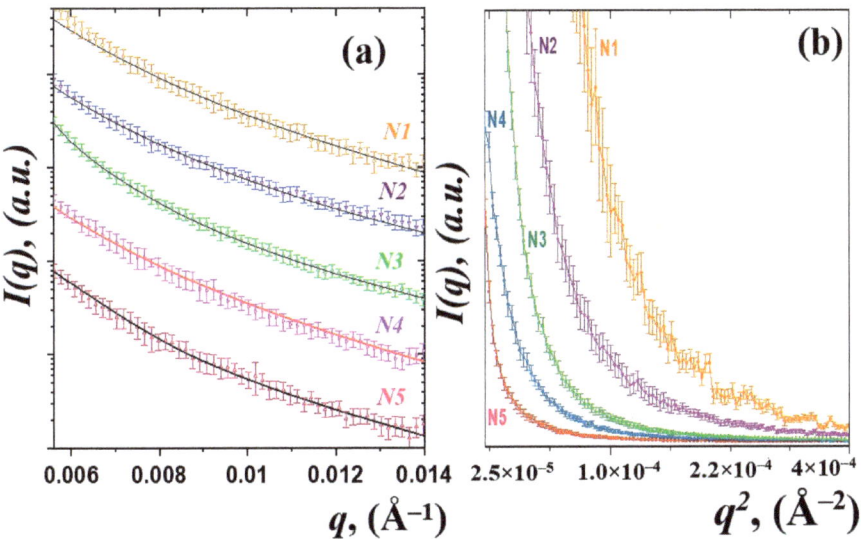

Figure 1. SAXS curves of the studied glass materials and their approximation by function (**a**). Guinier plots for the SAXS experimental data (**b**).

Therefore, the obtained SAXS curves were approximated by using the exponential-power law model of Beaucage [29,30]. Those approaches to the analysis of small-angle scattering describe scattering from complex systems that contain multiple levels of related structural features. It should be noted that even in the absence of nanoparticles in the glass matrix, glass density heterogeneities of different natures can act as scattering objects. The scattering intensity from a system of two scatters is represented by the following expression:

$$I(q) = G_1 \exp\left(\frac{-q^2 R_{g1}^2}{3}\right) + B_1 \exp\left(\frac{-q^2 R_{g1}^2}{3}\right)\left(\frac{1}{q_1^*}\right)^{P_1} + G_2 \exp\left(\frac{-q^2 R_{g2}^2}{3}\right) + B_2 \exp\left(\frac{-q^2 R_{g2}^2}{3}\right)\left(\frac{1}{q_2^*}\right)^{P_2}, \quad (1)$$

where the coefficients G_1, G_2, B_1 and B_2 and the degrees at exponents P_1 and P_2 are the fitted parameters for the first and second structural levels, respectively. The radius of gyration R_{g1} and R_{g2} correspond to the main parameters of the sizes of scattering objects. The functions q_1^* and q_2^* in a power function are normalized as:

$$q_1^* = \frac{q}{\left[\mathrm{erf}\left(\frac{k_1 q R_{g1}}{\sqrt{6}}\right)\right]^3}, \; q_2^* = \frac{q}{\left[\mathrm{erf}\left(\frac{k_2 q R_{g2}}{\sqrt{6}}\right)\right]^3} \quad (2)$$

where k_1 and k_2 are empirical coefficients. The values of the gyration radius R_{g1} and R_{g2} are correlated with fluctuations in the density of glass [10] and with the PbF$_2$:Tm-Ho nanoparticles, respectively.

The calculated values of the power-law exponents P_1 and P_2 obtained from fitting SAXS data by using Equation (1) are associated with the fractal dimension of the nanostructured system [24,31]. A power-law exponent in the range between 1 and 3 corresponds to mass fractals [24,25], one between 3 and 4 indicates surface fractals and between 4 and 6 is a diffuse surface. It is evident that the fractal dimensions of the observed nanoparticles PbF$_2$:Tm-Ho vary slightly within the range $P_1 = 3.0 \div 3.6$, which can correspond to some estimated nanoparticles with a smooth, sharp interface [24,31]. At the same time, the slope degree P_2 of the SAXS curve related to the density fluctuations does not exceed 3, which corresponds to the mass fractals. Large regions of density fluctuations of the glass material are formed in the glass matrix and are governed by the essential features of the chemical interaction of the glass components [10,14]. The observed density fluctuations of the glass material can serve as nucleation centers [15] for the nanostructured particles PbF$_2$:Tm-Ho.

The results of the approximation of the experimental data by the function (1) are shown in Figure 1a. The calculated size variation of the density fluctuations and formed nanoparticles is shown in Figure 2a. It can be noted that, in heat-treated glasses doped with thulium and holmium ions, the nanoparticles of 42–46 nm in size (Figure 2b) are formed in a spherical approximation, where the diameter of the nanoparticles is calculated as $D = 2(5/3)^{1/2} R_g$ and the average sizes of density fluctuations in the glass matrix grow from 204(2) nm for sample N1 to 324(3) nm for sample N5.

Interestingly, as the relative concentration of Tm$_2$O$_3$/Ho$_2$O$_3$ oxides increases, both the average size of nanoparticle clusters and the glass density fluctuations are growing. It can be explained that rare-earth ions are localized not only in nanoparticles but also in the glass matrix in the form of oxides [5,14]. In order to estimate the average size ranges of both glass density fluctuations and nanoparticles, approximations of SAXS data using functions (1) and (2) were used. The SAXS technique is much superior when considering the determination of the size distribution on a several-nanometer length scale for opaque solutions and for solid specimens [31].

Scattering comprises not only contributions from the regularity of the space-filling ordering of particles but also from a single particle. The particle scattering can be mathematically formulated depending on the type of particle shape. In block copolymer microdomain systems, the Gauss distribution of the particle size has been assumed [24,31]. Only recently has direct determination of the discrete size distribution been available by fitting the theoretical scattering function to the experimentally obtained SAXS profile [27].

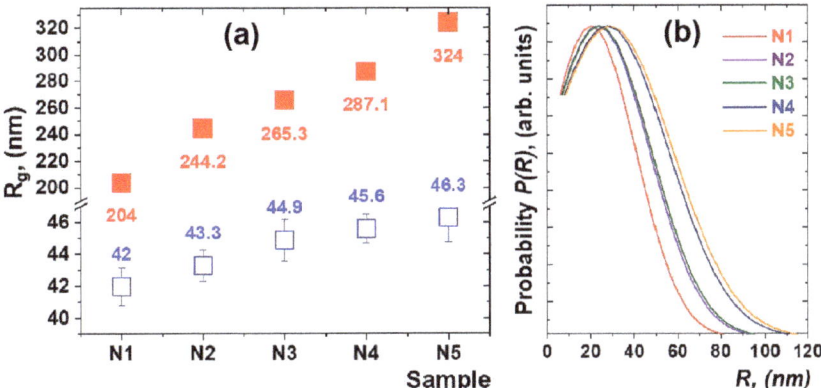

Figure 2. The values of the gyration radii Rg1 (red close squares) of the density fluctuations and Rg2 (blue open square) of nanoparticles, which were calculated from the analysis of the SAXS data for the studied glass materials. The obtained values are normalized to the corresponding values for N1 sample (**a**). The distribution of average sizes of nanoparticles in the studied glass samples obtained by Equations (3) and (4) (**b**).

The paired distribution function of nanoparticles of intermediate size, having a finite maximum size D_{max}, was approximated by a linear combination of a finite number of N cubic B-splines uniformly distributed in the range from 0 to D_{max}:

$$\rho(r) = \sum_{i=1}^{N} a_i \phi_i(r) \qquad (3)$$

where a_i is the coefficient of the i-th cubic B-spline, $\varphi_i(r)$ [31,32]. The upper limit of the values of the parameter r, included in the inverse Fourier transform, was chosen in such a way that the function $\rho(r)$ smoothly tends to 0 for large values of r. Using the above-mentioned mathematical apparatus, it is possible to estimate the paired distribution function for a system of non-interacting aggregates. Based on the obtained dependencies $\rho(r)$, it is possible to determine the radius of gyration Rg, which characterizes the size of intermediate nanoparticles:

$$R_g = \left[\frac{\int_0^D r^2 p(r) dr}{\int_0^D p(r) dr} \right]^{1/2} \qquad (4)$$

The results of the analysis are shown in Figure 2b. It can be seen that the average size of nanoparticles in the spherical approximation [31,32] shifts to the region of large sizes, although the width of the distribution does not change significantly. Interestingly, the slope of the SAXS curves practically does not change, and its average value is $\alpha = -5.1(5)$. This indicates that the fractal dimensionality and morphology of the nanostructured components of the heat-treated glasses are preserved.

As an important aspect, the structural mechanisms of PbF_2:Tm-Ho nanoparticle formation can also be explained by the detection of the crystalline or amorphous state of the luminescent nanoparticles. X-ray diffraction patterns for the studied glass samples are shown in Figure 3. All diffraction patterns obtained have a typical shape for scattering from amorphous materials. However, on the X-ray diffraction pattern corresponding to samples N2, N3 and N4, the appearance of several diffraction peaks is observed. The positions of the observed diffraction peaks correspond to the cubic structure with $Fm\overline{3}m$ symmetry and indicate the PbF_2 phase [5,14]. We believe that the rare-earth ions became embedded in crystals of PbF_2 because the unit cell parameter of this phase changes slightly with increasing thulium and holmium oxide concentration, which indicates the entry of Ho^{3+}

and Tm^{3+} ions into the crystal structure of the luminescent nanoparticle PbF_2:Tm-Ho. Here, it is declared that the spectral characteristics of up-conversion luminescence correspond to those related to the cubic crystal structure of PbF_2 crystal [4,33].

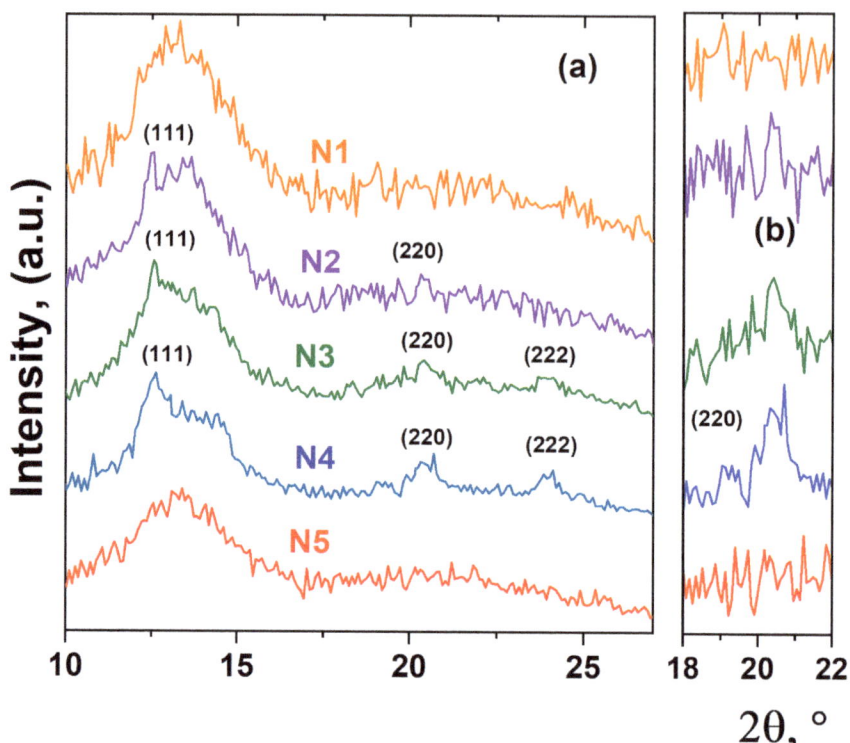

Figure 3. X-ray diffraction patterns of heat-treated glasses. The diffraction peaks of cubic phase PbF_2:Tm-Ho are indicated by Miller indices (**a**). The enlarged section of the diffraction pattern in scattering angle range 18–22°, where the diffraction reflex (220) of the cubic phase PbF_2:Tm-Ho is detected (**b**).

Based on the obtained experimental data, the following structural mechanism of nanoparticle formation in the heat-treated glass can be proposed. As previously assumed [5,10,19], the density fluctuations in the glass materials can serve as the nucleation centers for the oxide nanoparticles PbF_2:Tm-Ho. At low concentrations of the initial oxides Tm_2O_3 and Ho_2O_3, complex amorphous nanostructured structures, or aggregates, are formed. The nanoparticles form complex branching structures consisting of regular fractal arrangements of clusters inside the glass material. With increasing oxide concentration, the formation of a crystalline phase of PbF_2 nanoparticles with changes in the local environment of the glass matrix is observed. These crystallized PbF_2 nanoparticles are a host system for rare-earth Tm^{3+} and Ho^{3+} ions, whose entry into the cubic crystal lattice of PbF_2 provides conditions for up-conversion luminescence [4,5].

4. Conclusions

The structural features of nanoparticle formation in heat-treated mixed oxyfluoride glasses have been studied using small-angle X-ray scattering and X-ray diffraction methods. It has been established that nanoparticles, presumably PbF_2:Tm-Ho with sizes of 42–46 nm, are formed at the selected heat-treatment mode. An increase in the average size of the density fluctuations in glass from 204(2) to 324(3) nm is observed. With an increase in the concentration of Ho_2O_3 and Tm_2O_3 oxides, the average sizes of nanoparticles and the sizes of local areas of density fluctuations both increased. The obtained structural information will be useful for the analysis of the optical properties of nanostructured up-conversion-luminescent glass ceramics.

Author Contributions: Conceptualization, E.T. and S.D.; methodology, S.D. and K.P.; formal analysis, S.D.; investigation, S.D. and K.P.; data curation, S.D. and K.P.; writing—original draft preparation, S.K. and E.T.; writing—review and editing, S.K. and E.T.; visualization, S.D. and S.K.; supervision, D.K.; project administration, D.K. and E.T. All authors have read and agreed to the published version of the manuscript.

Funding: This research did not receive any specific grant from funding agencies in the public, commercial, or not-for-profit sectors.

Informed Consent Statement: Not applicable.

Data Availability Statement: Not applicable.

Conflicts of Interest: The authors declare no conflict of interest.

References

1. Probst, J.; Dembski, S.; Milde, M.; Rupp, S. Luminescent nanoparticles and their use for in vitro and in vivo diagnostics. *Expert Rev. Mol. Diagn.* **2012**, *12*, 49–64. [CrossRef] [PubMed]
2. Fares, H.; Elhouichet, H.; Gelloz, B.; Férid, M. Silver nanoparticles enhanced luminescence properties of Er^{3+} doped tellurite glasses: Effect of heat treatmen. *J. Appl. Phys.* **2014**, *116*, 123504. [CrossRef]
3. Tikhomirov, V.K.; Furniss, D.; Reaney, I.M.; Beggiora, M.; Ferrari, M.; Montagna, M.; Rolli, R. Fabrication and characterization of nanoscale, Er^{3+}-doped, ultratransparent oxyfluoride glass ceramic. *Appl. Phys. Lett.* **2002**, *81*, 1937–1939. [CrossRef]
4. Loiko, P.A.; Rachkovskaya, G.E.; Zakharevich, G.B.; Kornienko, A.A.; Dunina, E.B.; Yasukevich, A.S.; Yumashev, K.V. Cooperative up-conversion in Eu^{3+}, Yb^{3+}-doped SiO_2–PbO–PbF_2–CdF_2 oxyfluoride glass. *J. Non-Cryst. Solids* **2014**, *392–393*, 39–44. [CrossRef]
5. Kichanov, S.E.; Kozlenko, D.P.; Gorshkova, Y.E.; Rachkovskaya, G.E.; Zakharevich, G.B.; Savenko, B.N. Structural studies of nanoparticles doped with rare-earth ions in oxyfluoride lead-silicate glasses. *J. Nanopart. Res.* **2018**, *20*, 54. [CrossRef]
6. Trusova, E.E.; Bobkova, N.M.; Gurin, V.S. Nature of color centers in silicate glasses with additions of cerium and titanium oxides. *Glass Ceram.* **2009**, *66*, 9–13. [CrossRef]
7. Bondar, I.V.; Gurin, V.S.; Solovey, N.P.; Molochko, A.P. Formation and optical properties of $CuInTe_2$ nanoparticles in silicate matrices. *Semiconductors* **2007**, *41*, 939–945.
8. Sigaev, V.N.; Golubev, N.V.; Usmanova, L.Z.; Stefanovich, S.Y.; Pernice, P.; Fanelli, E.; Aronne, A.; Champagnon, B.; Califano, V.; Vouagner, D.; et al. On the nature of the second-order optical nonlinearity of nanoinhomogeneous glasses in the Li_2O-Nb_2O_5-SiO_2 system. *Glass Phys. Chem.* **2006**, *33*, 97–105. [CrossRef]
9. Zhang, Y.; Wang, Y. Nonlinear optical properties of metal nanoparticles: A review. *RSC Adv.* **2017**, *7*, 45129–45144. [CrossRef]
10. Kichanov, S.E.; Islamov, A.K.; Samoilenko, S.A.; Kozlenko, D.P.; Belushkin, A.V.; Gurin, V.S.; Shevchenko, G.P.; Trusova, E.E.; Bulavin, L.A.; Savenko, B.N. Studying the structural features of oxide nanoclusters of cerium and titanium in a silicate glass by means of the small-angle neutron scattering. *J. Surf. Investig.* **2014**, *8*, 98–103. [CrossRef]
11. Samoylenko, S.A.; Kichanov, S.E.; Belushkin, A.V.; Kozlenko, D.P.; Garamus, V.M.; Gurin, V.S.; Trusova, E.A.; Shevchenko, G.P.; Rakhmanov, S.K.; Bulavin, L.A.; et al. Study of Structural Aspects of the Cluster Formation in Silicate Glasses Doped with Cerium and Titanium Oxides by Small-Angle Neutron Scatterin. *Phys. Solid State* **2011**, *53*, 2431–2434. [CrossRef]
12. Godard, A. Infrared (2–12 μm) solid-state laser sources: A review. *Comptes Rendus Phys.* **2007**, *8*, 1100–1128. [CrossRef]
13. Nishibu, S.; Nishio, T.; Yonezawa, S.; Takashima, M. Fluorescence enhancement of oxide fluoride glass co-doped with TbF_3 and SmF_3. *J. Lumin.* **2007**, *126*, 365–370. [CrossRef]
14. Kichanov, S.E.; Gorshkova, Yu.E.; Rachkovskay, G.E.; Kozlenko, D.P.; Zakharevich, G.B.; Savenko, B.N. Structural evolution of luminescence nanoparticles with rare-earth ions in the oxyfluoride glass ceramics. *Mater. Chem. Phys.* **2019**, *237*, 121830–121837. [CrossRef]
15. Golubkov, V.V.; Bogdanov, V.N.; Pakhnin, A.Y.; Solovyev, V.A.; Zhivaeva, E.V.; Kabanov, V.O.; Yanush, O.V.; Nemilov, S.V.; Kisliuk, A.; Soltwisch, M.; et al. Microinhomogeneities of glasses of the system PbO–SiO_2. *J. Chem. Phys.* **1999**, *110*, 4897–4906. [CrossRef]
16. Auzel, F. Up-conversion and anti-Stokes processes with d and f ions in solids. *Chem. Rev.* **2004**, *104*, 139–174. [CrossRef]
17. Loiko, P.A.; Rachkovskaya, G.E.; Zakharevich, G.B.; Skoptsov, N.A.; Yumashev, K.V. Luminescence of Oxyfluoride Glasses Containing Yb^{3+}–RE^{3+} Ions. *Glass Ceram.* **2016**, *73*, 9–13. [CrossRef]
18. Guinhos, F.C.; Nóbrega, P.C.; Santa-Cruz, P.A. Compositional dependence of up-conversion process in Tm^{3+}–Yb^{3+} codoped oxyfluoride glasses and glass-ceramics. *J. Alloys Compd.* **2001**, *323–324*, 358–361. [CrossRef]

19. Dwivedi, Y.; Thakur, S.N.; Rai, S.B. Study of frequency upconversion in Yb^{3+}/Eu^{3+} by cooperative energy transfer in oxyfluoroborate glass matrix. *Appl. Phys.* **2007**, *B 89*, 45–51. [CrossRef]
20. Maciel, G.S.; Biswas, A.; Prasad, P.N. Infrared-to-visible $Eu^{3}+$ energy up-conversion due to cooperative energy transfer from an Yb^{3+} ion pair in a sol–gel processed multicomponent silica glass. *Opt. Commun.* **2000**, *178*, 65–69. [CrossRef]
21. Kochanowicz, M.; Zmojda, J.; Miluski, P.; Baranowska, A.; Leich, M.; Schwuchow, A.; Jäger, M.; Kuwik, M.; Pisarska, J.; Pisarsk, W.A.; et al. Tm^{3+}/Ho^{3+} co-doped germanate glass and double-clad optical fiber for broadband emission and lasing above 2 µm. *Opt. Mater. Express* **2019**, *9*, 1450–1458. [CrossRef]
22. Dwaraka Viswanath, C.S.; Babu, P.; Martín, I.R.; Venkatramu, V.; Lavín, V.; Jayasankar, C.K. Near-infrared and upconversion luminescence of Tm^{3+} and Tm^{3+}/Yb^{3+}-doped oxyfluorosilicate glasses. *J. Non-Cryst. Solids.* **2018**, *507*, 1–10. [CrossRef]
23. Richards, B.; Shen, S.; Jha, A.; Tsang, Y.; Binks, D. Infrared emission and energy transfer in Tm3+, Tm3+-Ho3+ and Tm^{3+}-Yb^{3+}-doped tellurite fibre. *Opt. Express.* **2007**, *15*, 6546–6551. [CrossRef]
24. Brumberger, H. *Modern Aspects of Small-Angle Scattering*; Springer: Dordrecht, The Netherlands, 1995. [CrossRef]
25. Svergun, D.I. Determination of the regularization parameter in indirect-transform methods using perceptual criteria. *J. Appl. Cryst.* **1992**, *25*, 495–503. [CrossRef]
26. Röding, M.; Tomaszewski, P.; Yu, S.; Borg, M.; Rönnols, J. Machine learning-accelerated small-angle X-ray scattering analysis of disordered two- and three-phase materials. *Front. Mater.* **2022**, *9*, 956839–956852. [CrossRef]
27. SasView for Small-Angle Scattering Analysis. Available online: http://www.sasview.org/ (accessed on 20 May 2019).
28. Rutkauskas, A.V.; Gorshkova, Y.E.; Gurin, V.S.; Kichanov, S.E.; Kozlenko, D.P.; Alekseenko, A.A. Investigation of Silicate Sol–Gel Glass Doped with Cu2Se and Eu Nanoparticles by Small-Angle Neutron Scattering and Atomic-Force Microscopy. *J. Surf. Investig. X-Ray Synchrotron Neutron Tech.* **2022**, *16*, 1094–1100. [CrossRef]
29. Beaucage, G. Approximations leading to a unified exponential/power-law approach to small-angle scattering. *J. Appl. Cryst.* **1995**, *28*, 71–728. [CrossRef]
30. Hammouda, B. Analysis of the Beaucage model. *J. Appl. Cryst.* **2010**, *43*, 1474–1478. [CrossRef]
31. Teixeira, J. Small-angle scattering by fractal systems. *J. Appl. Cryst.* **1988**, *21*, 781–785. [CrossRef]
32. Schmidt, P.W. Small-angle scattering studies of disordered, porous and fractal systems. *J. Appl. Cryst.* **1991**, *24*, 414–435. [CrossRef]
33. Bevan, D.J.M.; Strähle, J.; Greis, O. The crystal-structure of tveitite, an ordered yttrofluorite mineral. *J. Solid State Chem.* **1982**, *44*, 75–81. [CrossRef]

Disclaimer/Publisher's Note: The statements, opinions and data contained in all publications are solely those of the individual author(s) and contributor(s) and not of MDPI and/or the editor(s). MDPI and/or the editor(s) disclaim responsibility for any injury to people or property resulting from any ideas, methods, instructions or products referred to in the content.

Article

Barium Silicate Glasses and Glass–Ceramic Seals for YSZ-Based Electrochemical Devices

Alyona Vepreva [1], Dmitry Dubovtsev [1], Daria Krainova [1], Yulia Chetvertnykh [1], Semyon Belyakov [2], Nailya Saetova [1,3] and Anton Kuzmin [1,3,*]

[1] Institute of Chemistry and Ecology, Vyatka State University, Kirov 610000, Russia; a.vepreva98@mail.ru (A.V.); d.dubovtzev@yandex.ru (D.D.); dashakraynova@yandex.ru (D.K.); usr22050@vyatsu.ru (Y.C.); n.saetova@yandex.ru (N.S.)
[2] Laboratory of Kinetics, Institute of High-Temperature Electrochemistry, Ural Branch of Russian Academy of Sciences, Yekaterinburg 620137, Russia; bca2@mail.ru
[3] Laboratory of Solid State Chemistry, Institute of Solid State Chemistry and Mechanochemistry, Siberian Branch of Russian Academy of Sciences, Novosibirsk 630128, Russia
* Correspondence: a.v.kuzmin@yandex.ru

Abstract: The effect of partial SiO_2 substitution with Al_2O_3 and B_2O_3 on the thermal properties and crystallization of glass sealants in the $(50 - x)SiO_2$–$30BaO$–$20MgO$–$xAl_2O_3(B_2O_3)$ (wt %) system is studied. It is established that the coefficient of thermal expansion of all obtained glasses lies within a range of 8.2–9.9×10^{-6} K^{-1}. Alumina-doped glasses crystallize after quenching, while samples containing boron oxide are completely amorphous. Magnesium silicates are formed in all glasses after exposure at 1000 °C for 125 h. After 500 h of exposure, a noticeable diffusion of zirconium ions is observed from the YSZ electrolyte to the glass sealant volume, resulting in the formation of the $BaZrSi_3O_9$ compound. The crystallization and products of interaction between YSZ ceramics and boron-containing sealants have no significant effects on the adhesion and properties of glass sealants, which makes them promising for applications in electrochemical devices.

Keywords: glass; glass–ceramic; sealant; yttria-stabilized zirconia; microstructure; crystallization; SOFC

1. Introduction

Yttria-stabilized zirconia (YSZ) is widely used for various high-temperature devices, including gas sensors [1,2], fuel cells [3], and electrolyzers [4]. YSZ-based gas sensors are applied in various fields from medicine to the control of vehicle exhaust emissions [5] due to the high sensitivity to such gases as NO_x, CO, H_2, and hydrocarbons [6]. High-temperature annealing is frequently used to combine the sensor's parts [7] and create tight contact. Another approach to sensor assembling implies the application of inorganic binders, mainly a mixture of liquid glass and alumina powder [5,8]. Despite the fact that glass sealants are the least spread for electrochemical sensors, there are some studies indicating the perspective of such an approach [9]. In addition, glass sealants are the most suitable for sensors operating at high (above 1000 °C) temperatures, which is confirmed by the latest developments of the company Schott (Germany) (one of the largest sealant manufacturers on the international market), who presented a high-temperature sensor for monitoring the composition of car exhaust gases, assembled using a glass–ceramic sealant (data on the glass–ceramic high-temperature sealant HEATEN produced by Schott (Germany) can be found at https://www.schott.com/en-gb/products/heatan-p1000279/technical-details?tab= e5001c8e5b8b497997de6e65e33174f5, accessed on 21 June 2023).

As mentioned above, YSZ is widely used for a number of high-temperature devices, including oxygen pumps [10,11]. Oxygen pumps require tight sealing that can be reached by glass application [12–14]. In addition, high-temperature glass and glass–ceramic sealants are required to seal laboratory cells when conducting different experiments [15,16]. In addition to oxygen pumps, solid oxide fuel cells (SOFCs) are one of the common high-temperature devices widely using glasses and glass–ceramics as sealing materials [3].

Barium silicate glasses are the most widespread glass and glass–ceramic sealants due to their good stability at high temperatures, high mechanical strength, appropriate coefficient of thermal expansion, and low electrical conductivity [17–19]. The properties of glass sealants can be controlled by changing the glass composition [20–22] and the introduction of filles into the glass matrix to obtain composites [19,23,24]. Several scientific groups studying barium-containing and barium-free glass sealants and led by the following scientists can be mentioned: K. Singh and G. Kaur [20,25,26], F. Smeacetto and A.G. Sabato [27–29], X. Wang and Y. Dong [19,30,31], and A. Kuzmin and N. Saetova [32–35]. However, the sealant compositions presented in the cited studies are complex and contain more than four oxides. In this study, we aimed to develop glass sealants of less complex compositions containing three main oxides (SiO_2, BaO, and MgO) and small amounts of additives (to 4 wt %). In addition, the developed glass sealants are expected to be used not only for SOFC joints, but for sealing oxygen pumps, which can be sealed at higher temperatures than SOFCs.

The choice of glass components is conditioned by their role in the glass network and their effect on glass properties. Thus, SiO_2 is a glass former [22,36], and BaO and MgO are modifiers increasing glass transition and softening temperatures, which is vital for high-temperature applications and CTE value [22,36]. B_2O_3 (a glass-forming oxide) was added to improve wettability and suppress crystallization [20,36]. Al_2O_3, which can act as both a glass former and modifier depending on the concentration, was also introduced to suppress crystallization [20,36] and increase the long-term stability of the sealant [22,36]. The introduction of small amounts of additives is believed not to affect the glass properties dramatically (for instance, CTE and sealing temperature), but to impact the crystallization behavior of glasses and their stability under high temperatures. It should also be mentioned that, unlike other studies [37–40], the sealants under investigation contain both BaO and MgO.

This work is devoted to the investigation of the effect of the partial substitution of silica in the $(50 - x)SiO_2$–30BaO–20MgO–$xAl_2O_3(B_2O_3)$ (wt %) with Al_2O_3 and B_2O_3 on the thermal properties and crystallization of glass–ceramic sealants for high-temperature applications. The choice of glass composition is based on previous studies that utilized the $45SiO_2$–$15Al_2O_3$–25BaO–15MgO (wt %) glass for oxygen pump sealing [41].

2. Materials and Methods

$(50 - x)SiO_2$-30BaO-20MgO-xAl_2O_3/B_2O_3 (wt %) glasses were obtained by melting the stoichiometric mixtures of SiO_2, $BaCO_3$, $MgCO_3$, Al_2O_3, and B_2O_3 (99.99% purity) in alundum crucibles at a temperature of 1500 °C, followed by pouring the melt into a glassy- carbon mold. Annealing was performed at a temperature of T_g—50 °C for 1 h, and then the glass was cooled naturally in a furnace to room temperature. The chemical composition of the obtained glasses was determined by atomic emission spectroscopy (AES) using an Optima 4300 DV (Perkin Elmer, Waltham, MA, USA) spectrometer with an accuracy of 2–3%. The phase composition of the glasses and glass–ceramics was studied by X-ray diffraction (XRD) using an XRD-7000 (Shimadzu, Kyoto, Japan) and a D/MAX-2200 (Rigaku, Tokyo, Japan) diffractometer with Cu-K_α (λ= 1.5418 Å) radiation. XRD patterns were collected at room temperature in a 2θ range from 10 to 80° with a scanning step of 2 °/min.

To study the thermal expansion of the obtained materials, samples were cut out of as-cast glasses and glass–ceramic samples were prepared by the compaction of glass powders followed by sintering at 1050 °C. The measurements were conducted in temperature ranges of 50–800 °C (cut samples) and 50–720 °C (compacted samples) using a quartz dilatometer with a TT-80 (Tesatronic, Renens, Switzerland) meter with an accuracy of 0.01 μm; the heating rate was 2 °/min.

Hot-stage microscopy (HSM) was applied to study the sealant behavior under heating by means of an ODP 868 (TA Instruments, New Castle, DE, USA) optical dilatometry platform; the measurements were conducted in a heating microscope mode with a rate of 2 °/min. Samples were obtained by the compaction of glass powders; YSZ ceramic was used as a substrate. The glass transition and crystallization temperatures were determined by differential scanning calorimetry (DSC) using a 449 F1 Jupiter (Netzsch, Selb, Germany) simultaneous thermal analysis device. The following measurement conditions were set: platinum crucibles, a temperature range of 35–1100 °C, air atmosphere, and a heating rate of 10 °/min.

The glass powder mixed with ethyl alcohol was applied onto the YSZ surface to study the behavior of the sealant in contact with joined materials. Then, the samples were heat treated by the sealing mode (temperature of 1240 °C, 10 min, heating rate of 2 °/min) in an oxidizing atmosphere and cooled to room temperature in a furnace. The morphology of YSZ–sealant–YSZ sealed samples was studied by scanning electron microscopy (SEM) and energy dispersive spectroscopy (EDS) using a JSM-6510 LV (JEOL, Tokyo, Japan) microscope equipped with an Inca Energy 350 (Oxford Instruments, Abingdon, UK) energy dispersive spectroscopy system with an X-max 80 detector. Cross sections of samples were obtained by epoxy impregnation, followed by grinding and polishing using a P12Sb (Polilab, Moscow, Russia). SEM images were obtained in backscattered electron (BSE) mode to provide a contrast between the glass matrix and crystallized phases.

3. Results and Discussion

Table 1 presents the nominal glass compositions and those determined by AES. In general, the real compositions of the glasses are close to the nominal ones and the differences are within the method error. However, there is an interaction between alundum crucibles and glass melts, which is seen in the Al_2O_3 content.

Table 1. Nominal and real (AES) compositions of glasses in the $(50 - x)SiO_2$-$30BaO$-$20MgO$-$xAl_2O_3(B_2O_3)$ system (wt %).

Sample	SiO_2	MgO	BaO	Al_2O_3	B_2O_3
3B	47.0	20.0	30.0	-	3.0
3B AES	46.3	17.9	31.4	1.8	2.6
3A	47.0	20.0	30.0	3.0	-
3A AES	43.9	20.2	32.1	3.8	-
4B	46.0	20.0	30.0	-	4.0
4B AES	45.7	21.7	27.4	1.6	3.6
4A	46.0	20.0	30.0	4.0	-
4A AES	44.9	18.4	31.8	4.9	-

The XRD patterns given in Figure 1 demonstrate a broad halo typical of glasses. The appearance of a less pronounced halo near 40° in addition to the main halo observed in an angle range of ~20–30° could be connected with a phase separation [42].

After quenching, the samples doped with alumina demonstrated visible separation in the transparent (lower part) and opaque (upper part) layers, which is schematically demonstrated in Figure 1. To determine the phase composition of each part, the obtained glasses were cut, the separated parts were powdered, and XRD patterns were then collected (these patterns are given in Figure 1). According to the XRD patterns collected from the transparent and opaque regions of 4A glass (Figure 1), some XRD peaks indicating the presence of crystalline phases are seen in the opaque glass, while its transparent part is amorphous. The crystalline phase can be identified as magnesium silicate Mg_2SiO_4 (PDF card no. #078-1369). Glasses doped with boron oxide were homogeneous and transparent and demonstrated no visible phase separation or crystallization.

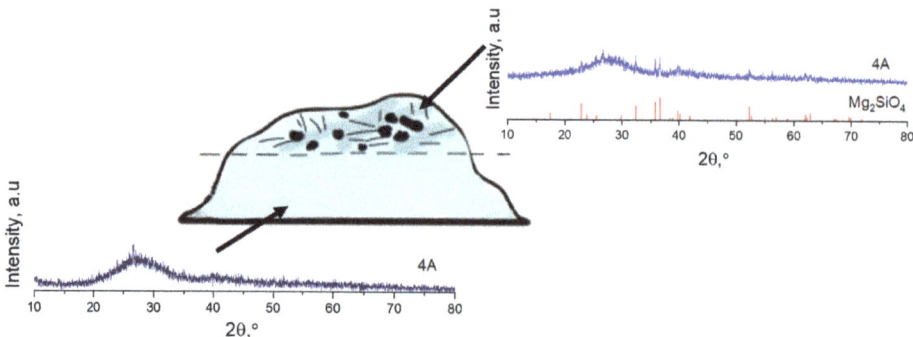

Figure 1. XRD patterns of as-cast 46SiO$_2$-30BaO-20MgO-4Al$_2$O$_3$ glass taken in different glass areas.

To obtain detailed information regarding phase separation in the 4A sample, its cross-section was studied by means of SEM and EDX. Figure 2a,b present SEM images with clearly pronounced areas in the glass volume consisting of needle-like crystals and broad light and dark bands. Since the images were obtained in the backscattered electron (BSE) mode, it can be assumed that the chemical composition of the mentioned regions differs, which is confirmed by the EDX mapping data. Since this method is insensitive to boron, some inaccuracies might appear during the EDX study. However, considering that the maximum boron content in the studied glasses is 4 wt % and EDX is the only method that can be used to characterize the phase composition of sealed glasses near the sealant–material interface, it was assumed that, in this case, boron could be excluded from further consideration without any significant loss in accuracy.

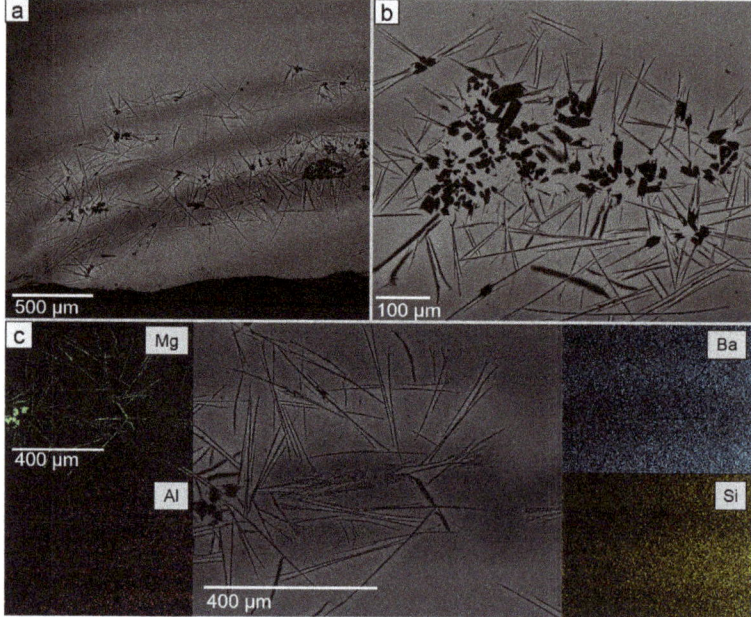

Figure 2. SEM images and element distribution maps for 4A as-cast glass: a – scale 500 μm, b – scale 100 μm, c – SEM images and element distribution maps. The scale of all maps is the same.

As seen in the element distribution maps given in Figure 2c, needle-like crystals can be attributed to manganese compounds, which correlated with XRD data (Figure 1), darker areas are enriched with silicon, and lighter areas possess an increased barium content. Thus, it is obvious that phase separation and crystallization occur during melt cooling after quenching. The boron-containing glasses do not show phase separation and crystallization after melt cooling.

When glasses are used as sealing materials, the thermal properties of the materials being joined and the operation temperatures must be considered. In this study, the sealants have been developed to join construction elements made of YSZ ceramics for which the value of the coefficient of thermal expansion (CTE) is ~9–10 × 10^{-6} K^{-1} [32]; this ceramic is used for oxygen pumps and sensors operating at rather high temperatures up to 1100 °C. The thermal characteristics of the glasses determined by DSC, dilatometry, and hot-stage microscopy are given in Table 2. It should be mentioned that the softening temperature (T_s) determined by HSM is surprisingly high and its value is far beyond the range between T_g and T_c, which is typical of glasses [43]. This might be explained by the crystallization of the studied samples during slow heating (2 °/min), leading to shifting all characteristic temperatures towards high temperatures.

Table 2. Thermal properties of $(50 - x)SiO_2$-30BaO-20MgO-$xAl_2O_3(B_2O_3)$ (wt %) glasses.

Sample	HSM, °C (±10 °C)	DSC, °C (±2 °C)		CTE × 10^{-6} K^{-1} (±0.1)	
	T_S	T_g	T_c	Bulk	Pressed
3A	1170	740	915	8.4	9.4
4A	1130	740	930	8.7	8.2
3B	1150	725	910	8.3	9.5
4B	1200	720	925	9.5	9.9

T_S—softening temperature.

As seen in the DSC data, the glass transition temperatures (T_g) of the boron-containing glasses are lower than those of the glasses doped with alumina. An increase in the content of both boron and aluminum oxides results in a slight growth of the crystallization temperature (T_c). It should be mentioned that in all cases, the crystallization temperatures are significantly lower than presumable operating temperatures; therefore, intense crystallization might be expected during sealing and running.

Figure 3 presents the temperature dependences of the linear expansion of YSZ ceramics, as-cast glasses, and glass–ceramic samples obtained by pressing glass powders followed by sintering at 1050 °C. Typical dilatometric curves of as-cast glasses are shown in the example of 4A and 4B samples. Given that glasses usually soften at lower temperatures than glass–ceramics, the measurements of the glass samples were carried out in a narrower temperature range. In general, all studied glass sealants have good compatibility with YSZ ceramics in terms of thermal expansion. It should be mentioned that in the studied temperature range, a dome typical for the dilatometric curves of glasses is seen only for the 4B glass, which might be connected with the insufficient maximum temperature of the experiment or with the partial crystallization of the glasses during heating, which affects the curve shape. The values of CTE given in Table 2 were calculated in a temperature range of 50–500 °C. The CTE values of the glass–ceramic samples are slightly higher than those of glasses in most cases, except for the 4A composition. A greater CTE value of glass (8.7 × 10^{-6} K^{-1}) compared with glass–ceramics (8.2 × 10^{-6} K^{-1}) might be connected with the appearance of crystalline phases with a low CTE value during the glass–ceramics sample preparation.

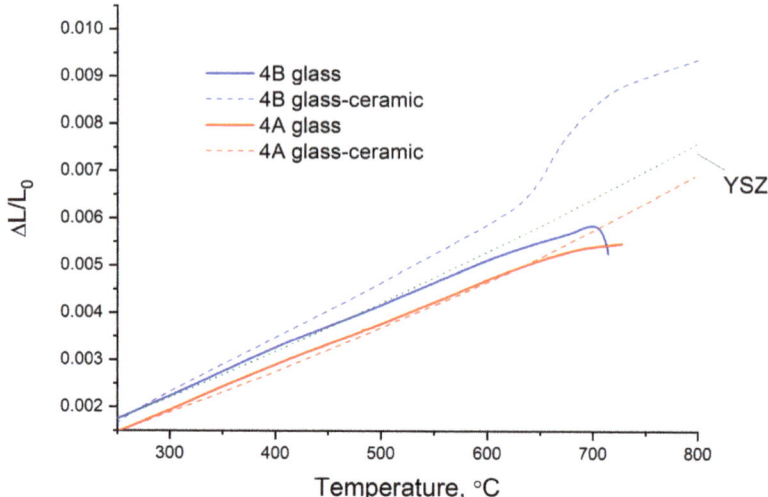

Figure 3. Curves of relative elongation of as-cast glasses, glass–ceramics, and YSZ ceramics.

According to XRD patterns collected after the sintering of pressed samples at 1050 °C (Figure 4), the 3B sample demonstrates the lowest tendency to crystallization: its crystallization degree was 15.1%, while that value was 79.2, 62.3, and 69.7 for samples 3A, 4A, and 4B, respectively. The percentage of crystalline phases was roughly estimated using the method described by Pardo [44] with an accuracy of ±2.5%. According to the phase identification, $BaMg_2Si_2O_7$ (CTE~10×10^{-6} K^{-1}) [45], $MgSiO_3$, and Mg_2SiO_4 were found.

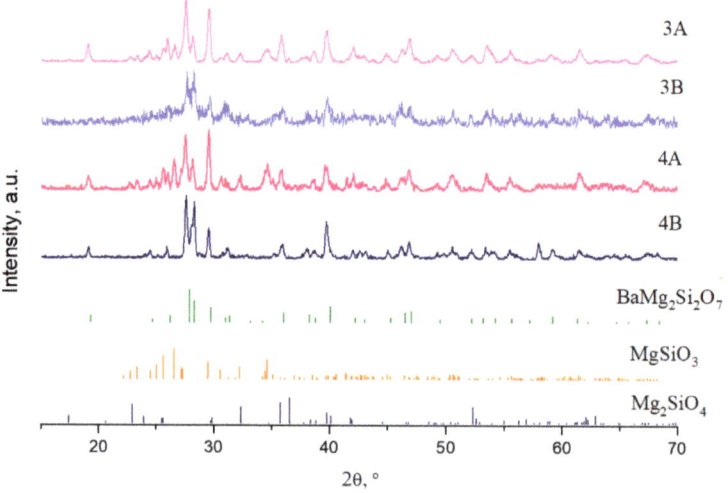

Figure 4. XRD patterns of $(50 - x)SiO_2$-30BaO-20MgO-xAl$_2$O$_3$(B$_2$O$_3$) glasses sintered at 1050 °C for 10 min. $BaMg_2Si_2O_7$ (PDF#10-0044), $MgSiO_3$ (PDF#018-0778), Mg_2SiO_4 (PDF#078-1369).

The behavior of the glasses under heating was studied using hot-stage microscopy, allowing the tracking of sintering, softening, sphere formation, and melting [46]. However, only the sintering temperature can be clearly determined for the studied samples, while further shape change is less pronounced. As seen in the HSM images given in Figure 5, the formation of a sphere typical of glasses [43,47,48] was not observed for some samples, and a shape change corresponding to melting appears after softening. This might be connected with the fact that the softening temperature (1140–1200 °C) is higher than the crystallization temperature (Table 2) and glass–ceramic samples with a high crystallinity degree are formed, which affects the sample behavior under heating.

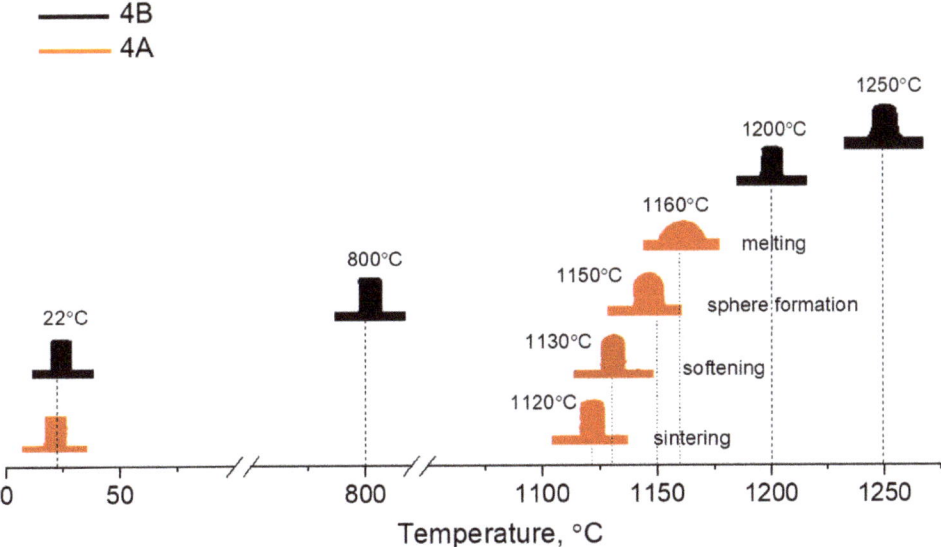

Figure 5. Hot-stage microscopy images for 4B and 4A samples.

As a rule, the sealing temperature is chosen based on the hot-stage microscopy data (when this method is used), and two approaches to its choice can be considered. If the sphere formation temperature is chosen as the sealing temperature, the sealing time is increased; if the half-sphere formation temperature is chosen as the sealing one, the sealing time is reduced [49]. However, this is impossible for the glasses under investigation due to intense crystallization causing the unusual behavior described above. Therefore, a temperature of 1240 °C was chosen empirically (based on the results of the experiment consisting of the measurement of wettability angles using cross-sections of YSZ–sealant samples [50]); the same sealing temperature was used for all glasses to reliably compare their crystallization during heat treatment.

Figure 6 presents the typical XRD patterns of YSZ–sealant on the examples of samples containing 3 wt % Al_2O_3 and B_2O_3. The samples were treated by the sealing mode: heating to 1240 °C with a rate of 2 °/min, exposing for 10 min, and natural cooling to room temperature. Then, the samples were put in a furnace, heated to 1000 °C, and kept for 125, 500, and 1000 h. As is seen, the main crystallization occurred during 125 h of exposure, which is typical for glasses [28,51,52], and further exposure did not lead to noticeable changes in the XRD patterns. The XRD peaks were attributed to enstatite ($MgSiO_3$), forsterite (Mg_2SiO_4), and YSZ substrate (30, 50, and 60°). It is worth noting that the $BaMg_2Si_2O_7$ compound found after sample sintering for CTE measurements (Figure 4) was not observed, which might be connected with its instability under sealing conditions. It should be noted that barium silicate glasses tend to the formation of numerous phases during crystallization, which could undergo phase transitions [51,53,54]. Therefore, the phase identification in such glasses using only XRD analysis is complicated.

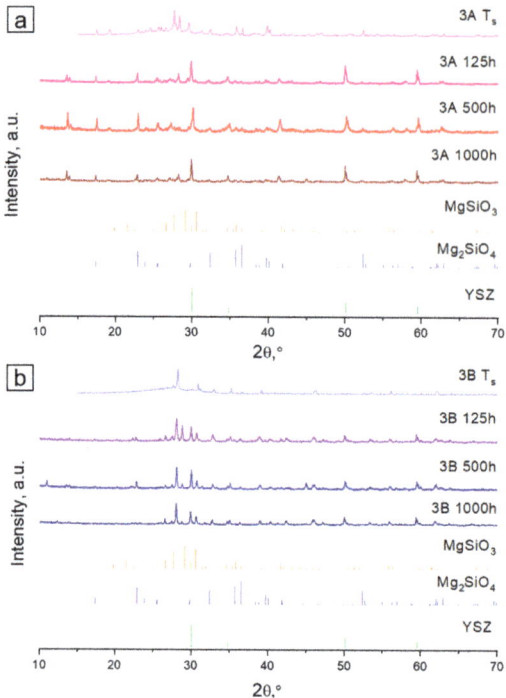

Figure 6. XRD patterns of YSZ–sealant samples of 3A (**a**) and 3B (**b**) glasses after heat treatment by the sealing mode and exposure for 125, 500, and 1000 h at 1000 °C in air atmosphere. $MgSiO_3$ (PDF#018-0778), Mg_2SiO_4 (PDF#078-1369).

To study the chemical interaction and crystallization processes, cross-sections of YSZ–sealant samples were prepared. The microstructure was studied by scanning electron microscopy in BSE mode because it is sensitive to the atomic weight of elements, and the crystalline phases of different compositions can be distinguished by contrast.

Figure 7 shows the SEM images and element distribution maps of YSZ–4A and YSZ–4B samples exposed at 1000 °C for 125 h. The observed inhomogeneity of the element distribution is caused by the crystallization and phase separation typical for barium-containing glasses [55] and demonstrated above for alumina-doped samples (Figure 2). In the presented SEM images, phase separation in glasses is seen from a slight difference in gray shadows: darker irregular areas can be distinguished in the light-gray glass (examples of such areas are highlighted in Figure 7). Thus, the darker area near point 1 (Figure 7a) is enriched with alumina (Table S1) while, the lighter area near point 3 is enriched with barium. It was also found that in boron-containing samples 3B and 4B, $MgSiO_3$ is formed (Figure 7a, point 2), while both Mg_2SiO_4 and $MgSiO_3$ appear in alumina-containing glasses (Figure 7a, spectrum 2 (Table S1) and Figure 8, spectra 6 and 7, Table S1).

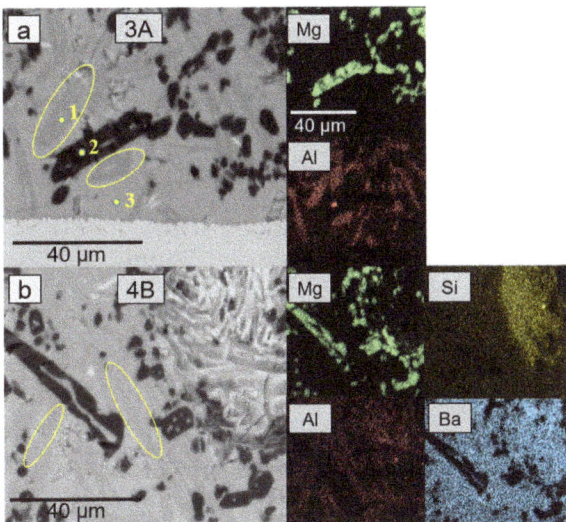

Figure 7. SEM images and element distribution maps of YSZ–3A (**a**) and YSZ–4B (**b**) joints after exposure at 1000 °C for 125 h in air atmosphere. Points 1–3 indicate areas of EDX study (chemical compositions are given in Table S1). Yellow circles indicate areas with increased alumina content. The scale of all element distribution maps is similar.

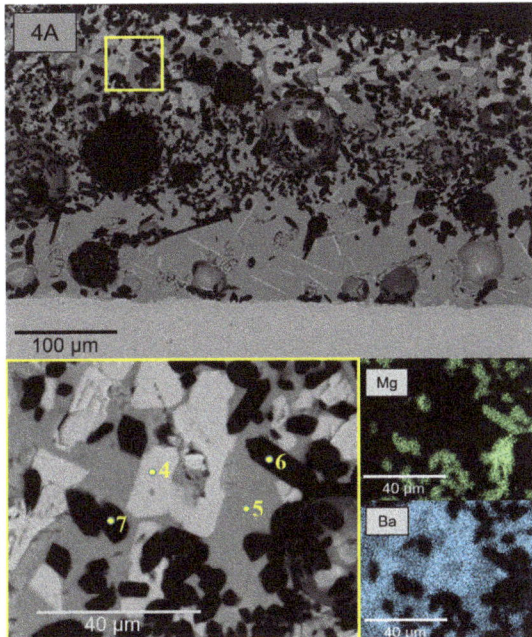

Figure 8. SEM images and element distribution maps of YSZ–4A joint after exposure at 1000 °C for 125 h in air atmosphere. Points 4–7 indicate areas of EDX study (chemical compositions are given in Table S1). Yellow square indicates the area of the upper SEM image given in higher magnification.

After exposure of the YSZ–4A sample for 125 h (Figure 8), areas with different chemical compositions are observed in BSE SEM images. For this composition, uneven phase distribution is typical: the composition of lighter areas (Figure 8, spectrum 1) is close to the $BaSi_2O_5$ compound (which can be both amorphous [56] and crystalline) and these areas are located near the sealant surface. The composition of the dark crystals of similar shape (spectra 6 and 7, Table S1) is close to the $MgSiO_3$ phase. It should be noted that no alumina-containing phases were found, which might indicate that it does not contribute to the phase formation. However, it should be mentioned that since the alumina content in the studied glass is low (4 wt %), aluminum-containing crystalline phases could be distributed unevenly and its presence might be missed due to the SEM limitations connected with the visible area. A thin uniform layer and needle-like crystals are observed near the sealant–YSZ interface, but their composition cannot be determined by EDX due to the small size.

Figure 9 shows SEM images of YSZ–sealant cross-sections after 500 h exposure at 1000 °C. Triangle-shaped light inclusions are observed in 3B and 4B sealants (Figure 9a, point 1) in addition to magnesium silicate crystals observed after 125 h exposure. Using EDX data, it was established that the chemical composition in point 1 (Figure 9) is close to the $BaZrSi_3O_9$ compound (experimental and theoretical compositions are given in Table 3). In addition, some amounts of yttria and zirconia were found in the glass volume (point 2). Obviously, both elements were transferred into the glass matrix due to ion diffusion from the YSZ ceramics during heat treatment. Although zirconium and yttrium diffusion was also observed in the case of the 3A and 4A sealants (Figure 9d, points 6 and 7), the formation of $BaZrSi_3O_9$ was not established. Some changes in magnesium silicate formation can be mentioned: while $MgSiO_3$ was formed in the 3B and 4B sealants boron-containing glasses after 125 h exposure, it was not found in the 3B sealant after 500 h exposure. Dark crystals seen in sealants 3A, 4A, and 3B (Figure 9, points 3 and 4) correspond to the Mg_2SiO_4 phase, while Mg_2SiO_3 is found in the 4B sample. According to the EDX analysis of residual glass (points 2, 5, 6, and 7), the glass matrix is depleted with magnesium and silicon, which is apparently caused by the intense crystallization of manganese silicates.

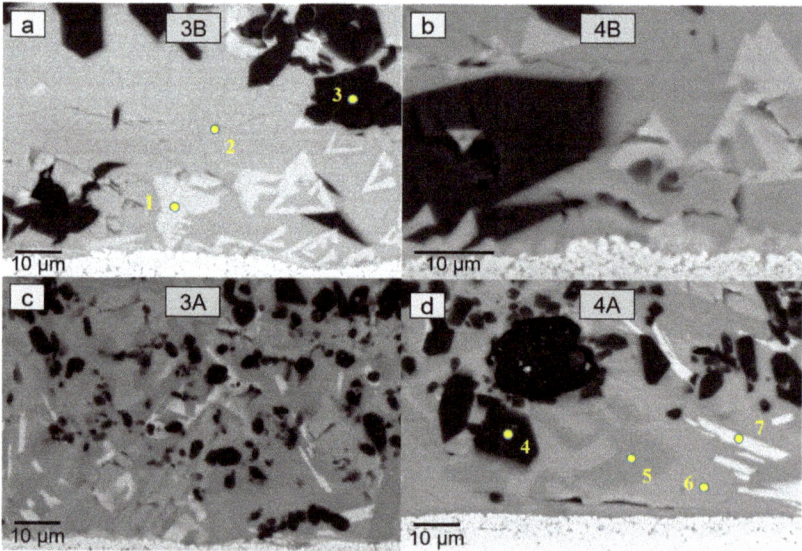

Figure 9. SEM images of sealant–YSZ interface after 500 h of exposure at 1000 °C in air atmosphere. Points correspond to the areas of EDX study: **a** – 3B sealant, **b** – 4B sealant, **c** – 3A sealant, and **d** – 4A sealant. Points 1–7 indicate areas of EDX study (chemical compositions are given in Table 3).

Table 3. Chemical composition determined by EDX in points depicted in Figure 9 (wt %) *.

Element (wt %)	Ba	Mg	Si	O	Al	Zr	Y
1	34.4	0.3	22.7	42.2	0.3	20.4	0.4
2	32.2	3.1	24.6	37.2	1.1	1.3	3.5
3	0.1	32.6	19.2	48.1	-	-	-
4	0.8	33.8	20.7	44.7	-	-	-
5	38.2	3.1	19.9	33.5	5.4	-	-
6	37.5	3.1	24.1	34.2	1.2	4.1	3.2
7	41.6	2.6	22.0	32.9	0.9	8.2	2.3
Mg_2SiO_4 **	-	34.6	20.0	45.5	-	-	-
$BaZrSi_3O_9$ **	30.1	-	18.4	31.5	-	20.0	-

*—excluding boron; **—theoretical values.

The further study of yttrium and zirconium diffusion to the glass volume was carried out using YSZ–sealant–YSZ joints kept at 1000 °C for 1000 h in an air atmosphere (Figure 10). According to the collected data, no new compounds were formed and the main crystalline phases are Mg_2SiO_4 and $BaZrSi_3O_9$ (Table S2). It is clearly seen that despite the similar phase composition, the distribution of the formed crystal over the glass volume differs for compositions substituted with boron and aluminum oxides. Thus, the magnesium silicate crystals formed in boron-containing glasses (Figure 10b) are larger compared with those in aluminum-containing glasses (Figure 10d), and are more evenly distributed over the glass volume. It should be noted that no $MgSiO_3$ crystals were observed after 1000 h exposure, which might be connected with the fact that, in terms of thermodynamics, Mg_2SiO_4 formation is preferable to $MgSiO_3$ formation [57–59]. As for the $BaZrSi_3O_9$ phase, it was only found in boron-containing glasses (Figure 10a,b). Although the size of the needle-like crystals located near alumina-doped sealant–YSZ interface increased, it is still insufficient to determine their chemical compositions using EDX.

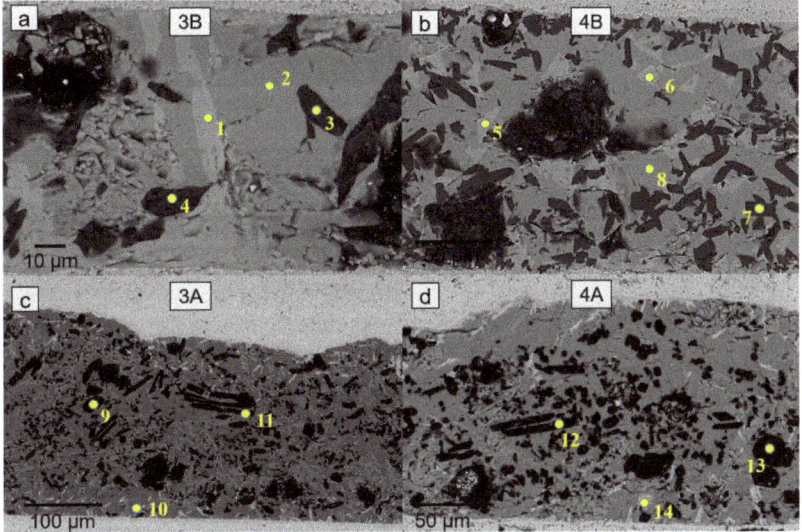

Figure 10. SEM images of YSZ–glass–YSZ joints after 1000 h exposure at 1000 °C in an air atmosphere: a – 3B sealant, b – 4B sealant, c – 3A sealant, and d – 4A sealant. Points correspond to the areas of EDX study; corresponding chemical compositions are given in Table S2.

To study the Zr^{2+} and Y^{3+} diffusion into the sealant volume, element distribution profiles were collected on the YSZ–3B (Figure 11a,c,d) samples after 125, 500, and 1000 h exposure at 1000 °C and the YSZ–3A (Figure 11b) sample after 1000 h exposure at 1000 °C in an air atmosphere. As is seen, despite the diffusion, zirconium is unevenly distributed over the glass volume, which can be explained by its binding into silicates during its interaction with the glass network. Obviously, the formation of the Zr-containing phase becomes more pronounced with an increase in the exposure time and the appearance of Zr-enriched regions is most clearly observed after 1000 h exposure. Nevertheless, some amount of yttrium and zirconium can be found in uncrystallized vitreous regions (Figure 10, Table S2) even in boron-free glasses, which allows one to expect Zr-containing phase formation with an increase in the exposure time.

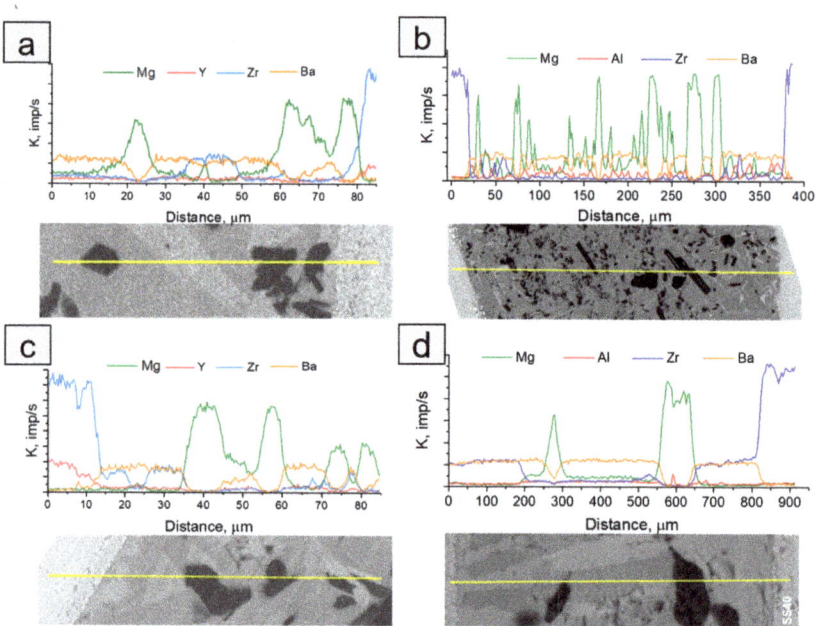

Figure 11. Element distribution over the YSZ–3B samples exposed for 125 (**a**), 500 (**c**), and 1000 h (**d**) at 1000 °C and YSZ–3A sample exposed for 1000 h (**b**).

Although Zr^{4+} and Y^{3+} diffusion into the sealant volume is well known for glasses containing barium and boron oxides [25,26,42,60], its mechanism has not been unambiguously explained yet. Moreover, less complex reaction products such as barium zirconate ($BaZrO_3$) [61] are typically formed. According to the $BaO-ZrO_2-SiO_2$ phase diagram [62], the $BaZrSi_3O_9$ compound can be obtained by the co-sintering of corresponding oxides at 1300 °C for 30 h, and it melts congruently at 1450 °C with the formation of $BaSi_2O_5$ and $ZrSiO_4$. However, there is some evidence that the $BaZrSi_3O_9$ phase can crystallize in glasses at temperatures below 1300 °C and much lower exposure times [63,64]. Preparing glass–ceramics using the unconventional solid-state method, Bo Li and co-authors suggest the following equation to describe the barium zirconium silicate formation: $BaO + ZrO_2 + 3SiO_2 = BaZrSi_3O_9$ [63,64], which is close to the one proposed by V. G. Chukhlantsev and Y. M. Galkin [62]. However, even though this equation appears to be suitable for ceramics and glass–ceramics obtained by sintering, it seems that it cannot be applied to describe the phase formation during glass crystallization due to the existence of an extended glass network.

In glasses similar to the ones studied in this work, BaO and SiO_2 can act as glass formers, MgO is a network modifier [65], and zirconium can be considered either as a glass former [66] or a modifier [67]. Therefore, it might be assumed that Zr^{4+} ions can embed into the Si-enriched part of the glass network forming $ZrSiO_4$ and then react with the Ba-enriched part of the glass network, which could lead to the $BaZrSi_3O_9$ formation. V. G. Chukhlantsev and Y. M. Galkin [62] proved the possibility of $BaZrSi_3O_9$ formation through the sintering of $BaSi_2O_5 + ZrO_2$ and $BaSi_2O_5 + ZrSiO_4$ mixtures at 1250 °C for 80 h. Since the sintering of the $BaSi_2O_5 + ZrO_2$ mixture results in the formation of $Ba_2Zr_2Si_3O_{12}$ composition, which was not observed in the studied glass, it could be assumed that the barium zirconium silicate formation in the glass network might be described by the $BaSi_2O_5 + ZrSiO_4 = BaZrSi_3O_9$ equation.

Zr^{4+} and Y^{3+} diffusion did not worsen the integrity of YSZ–sealant joints and it did not affect the adhesion to ceramics, which allows the sealants to be applied for the sealing of zirconia-based ceramics. However, yttrium diffusion from YSZ to glass could cause cubic → monoclinic transition in ZrO_2 [68] and the degradation of its surface, which might lead to joint failure.

All glass sealants passed the long exposure tests and showed good results in adhesion and thermal compatibility with the electrolyte. The most suitable option for further application is aluminum-containing glass because only a slight Zr^{4+} diffusion was observed and they can be expected to provide good sealing at a longer operating time. It should be mentioned that the behavior of the studied glasses differs from some glasses studied before. For example, the formation of $BaMg_2Si_2O_7$ in the phase observed in this work is hardly discussed in the literature because the formation of $Ba_2Si_3O_8$, $BaSi_2O_5$, and $BaAl_2Si_2O_8$ phases is more typical for barium-containing glasses [42,43,53].

4. Conclusions

Glasses with different Al_2O_3 and B_2O_3 content were obtained in the $(50 - x)SiO_2$-$30BaO$-$20MgO$-xAl_2O_3/B_2O_3 (wt.%) system. An increase in B_2O_3 content results in a decrease in the glass transition temperature, while the introduction of Al_2O_3 has the opposite effect. The CTE values of the samples do not depend on the composition, which may be due to crystallization upon heating. This ensures the long-term stability of the sealant–YSZ joints.

The crystallization processes in the studied glasses strongly depend on the heat treatment temperature. Heat treatment at 1050 °C results in the formation of $BaMg_2Si_2O_7$, $MgSiO_3$, and Mg_2SiO_4 phases in all of the studied glass sealants. Al_2O_3-containing glasses show a higher tendency to crystallize.

According to SEM and EDX studies of the behavior of the glass sealant in contact with YSZ ceramics, a higher tendency of Zr^{4+} to diffuse from the ceramic into the glass volume is observed for B_2O_3-containing glasses after exposure at 1000 °C. This diffusion results in the formation of the $BaZrSi_3O_9$ phase throughout the sealant. The Zr^{4+} diffusion depth in Al_2O_3-containing glasses is much smaller.

It can be concluded that the 4A composition is the most suitable for application in high-temperature devices based on YSZ ceramics, including solid oxide fuel cells, electrolyzers, and gas sensors. Despite a greater crystallization tendency, the presence of Al_2O_3 significantly inhibits the diffusion of Zr^{4+}, ensuring the stability of the sealant–YSZ joints. This sealant composition was successfully used to create an oxygen pump with a sensor that operates for a long time at temperatures of 850–1100 °C (Figure S1).

Supplementary Materials: The following supporting information can be downloaded at: https://www.mdpi.com/article/10.3390/ceramics6030081/s1: Table S1. EDX results for spectra provided in Figures 7 and 8. Table S2. EDX results for spectra provided in Figure 11. Figure S1. Oxygen pump with sensor glued with glass sealant

Author Contributions: Conceptualization, A.V. and N.S.; methodology, A.V., D.D. and D.K.; formal analysis, N.S. and S.B.; investigation, S.B., A.V., Y.C. and D.K.; data curation, N.S.; writing—original draft preparation, A.V., N.S. and D.K.; writing—review and editing, A.V. and N.S.; visualization, A.V.; supervision, A.K.; project administration, A.K. All authors have read and agreed to the published version of the manuscript.

Funding: The study of the interaction between glass and ceramics was funded by the Russian Science Foundation, grant number 21-79-30051.

Institutional Review Board Statement: Not applicable.

Informed Consent Statement: Not applicable.

Data Availability Statement: The data presented in this study are available on request from the corresponding author.

Acknowledgments: The research was partially performed using the facilities of the Shared Access Centre "Composition of Compounds" of IHTE UB RAS. The authors are grateful to A. V. Khodimchuk for her kind assistance with the XRD experiments.

Conflicts of Interest: The authors declare no conflict of interest.

References

1. Cirera, A.; Lpez-Gándara, C.; Ramos, F.M. YSZ-based oxygen sensors and the use of nanomaterials: A review from classical models to current trends. *J. Sens.* 2009, *2009*, 15. [CrossRef]
2. Muroyama, H.; Okuda, S.; Matsui, T.; Hashigami, S.; Kawano, M.; Inagaki, T.; Eguchi, K. Gas composition analysis using yttria-stabilized zirconia oxygen sensor during dry reforming and partial oxidation of methane. *J. Jpn. Pet. Inst.* 2018, *61*, 72–79. [CrossRef]
3. Liu, M.; Dong, D.; Peng, R.; Gao, J.; Diwu, J.; Liu, X.; Meng, G. YSZ-based SOFC with modified electrode/electrolyte interfaces for operating at temperature lower than 650 °C. *J. Power Sources* 2008, *180*, 215–220. [CrossRef]
4. Hauch, A.; Brodersen, K.; Chen, M.; Mogensen, M.B. Ni/YSZ electrodes structures optimized for increased electrolysis performance and durability. *Solid State Ion.* 2016, *293*, 27–36. [CrossRef]
5. Jiang, L.; Lv, S.; Tang, W.; Zhao, L.; Wang, C.; Wang, J.; Wang, T.; Guo, X.; Liu, F.; Wang, C.; et al. YSZ-based acetone sensor using a Cd2SnO4 sensing electrode for exhaled breath detection in medical diagnosis. *Sens. Actuators B Chem.* 2021, *345*, 130321. [CrossRef]
6. Miura, N.; Sato, T.; Anggraini, S.A.; Ikeda, H.; Zhuiykov, S. A review of mixed-potential type zirconia-based gas sensors. *Ionics* 2014, *20*, 901–925. [CrossRef]
7. Tho, N.D.; Van Huong, D.; Ngan, P.Q.; Thai, G.H.; Thu, D.T.A.; Thu, D.T.; Tuoi, N.T.M.; Toan, N.N.; Giang, H.T. Effect of sintering temperature of mixed potential sensor Pt/YSZ/LaFeO3 on gas sensing performance. *Sens. Actuators B Chem.* 2016, *224*, 747–754. [CrossRef]
8. Wang, J.; Wang, C.; Liu, A.; You, R.; Liu, F.; Li, S.; Zhao, L.; Jin, R.; He, J.; Yang, Z.; et al. High-response mixed-potential type planar YSZ-based NO2 sensor coupled with CoTiO3 sensing electrode. *Sens. Actuators B Chem.* 2019, *287*, 185–190. [CrossRef]
9. Schubert, F.; Wollenhaupt, S.; Kita, J.; Hagen, G.; Moos, R. Platform to develop exhaust gas sensors manufactured by glass-solder-supported joining of sintered yttria-stabilized zirconia. *J. Sens. Sens. Syst.* 2016, *5*, 25–32. [CrossRef]
10. Zhang, X.D.; Li, J.J.; Guo, X. Oxygen pump based on stabilized zirconia. *Rev. Sci. Instrum.* 2015, *86*, 115103. [CrossRef]
11. Tao, M.; Feng, J.; Li, R.; Guan, C.; Wang, J.; Chi, B.; Pu, J. Interfacial compatibility and thermal cycle stability for glass-sealed oxygen sensors. *Ceram. Int.* 2023, *49*, 23180–23188. [CrossRef]
12. Spirin, A.; Lipilin, A.; Ivanov, V.; Paranin, S.; Nikonov, A.; Khrustov, V.; Portnov, D.; Gavrilov, N.; Mamaev, A. Solid Oxide Electrolyte Based Oxygen Pump. In Proceedings of the 12th International Ceramics Congress Part D, Montecatini Terme, Italy, 6–11 June 2010; Volume 65, pp. 257–262.
13. Spirin, A.V.; Nikonov, A.V.; Lipilin, A.S.; Paranin, S.N.; Ivanov, V.V.; Khrustov, V.R.; Valentsev, A.V.; Krutikov, V.I. Electrochemical cell with solid oxide electrolyte and oxygen pump thereof. *Russ. J. Electrochem.* 2011, *47*, 569–578. [CrossRef]
14. Pham, A.Q.; Glass, R.S. Oxygen pumping characteristics of yttria-stabilized-zirconia. *Electrochim. Acta* 1998, *43*, 2699–2708. [CrossRef]
15. Khodimchuk, A.V.; Anan'ev, M.V.; Eremin, V.A.; Tropin, E.S.; Farlenkov, A.S.; Porotnikova, N.M.; Kurumchin, E.K.; Bronin, D.I. Oxygen isotope exchange between the gas-phase and the electrochemical cell O_2, Pt | YSZ | Pt, O_2 under conditions of applied potential difference. *Russ. J. Electrochem.* 2017, *53*, 838–845. [CrossRef]
16. Gunduz, S.; Dogu, D.; Deka, D.J.; Meyer, K.E.; Fuller, A.; Co, A.C.; Ozkan, U.S. Application of solid electrolyte cells in ion pump and electrolyzer modes to promote catalytic reactions: An overview. *Catal. Today* 2019, *323*, 3–13. [CrossRef]
17. Sohn, S.B.; Choi, S.Y.; Kim, G.H.; Song, H.S.; Kim, G.D. Suitable Glass-Ceramic Sealant for Planar Solid-Oxide Fuel Cells. *J. Am. Ceram. Soc.* 2004, *87*, 254–260. [CrossRef]
18. Batfalsky, P.; Haanappel, V.A.C.; Malzbender, J.; Menzler, N.H.; Shemet, V.; Vinke, I.C.; Steinbrech, R.W. Chemical interaction between glass-ceramic sealants and interconnect steels in SOFC stacks. *J. Power Sources* 2006, *155*, 128–137. [CrossRef]
19. Wang, X.; Yan, D.; Fang, D.; Luo, J.; Pu, J.; Chi, B.; Jian, L. Optimization of Al2O3-glass composite seals for planar intermediate-temperature solid oxide fuel cells. *J. Power Sources* 2013, *226*, 127–133. [CrossRef]
20. Singh, K.; Walia, T. Review on silicate and borosilicate-based glass sealants and their interaction with components of solid oxide fuel cell. *Int. J. Energy Res.* 2021, *45*, 20559–20582. [CrossRef]

21. Meinhardt, K.D.; Kim, D.S.; Chou, Y.S.; Weil, K.S. Synthesis and properties of a barium aluminosilicate solid oxide fuel cell glass-ceramic sealant. *J. Power Sources* **2008**, *182*, 188–196. [CrossRef]
22. Mahapatra, M.K.; Lu, K. Glass-based seals for solid oxide fuel and electrolyzer cells—A review. *Mater. Sci. Eng. R Reports* **2010**, *67*, 65–85. [CrossRef]
23. Kosiorek, M.; Żurawska, A.; Ajdys, L.; Kolasa, A.; Naumovich, Y.; Wiecińska, P.; Yaremchenko, A.; Kupecki, J. Glass–Zirconia Composites as Seals for Solid Oxide Cells: Preparation, Properties, and Stability over Repeated Thermal Cycles. *Materials* **2023**, *16*, 1634. [CrossRef] [PubMed]
24. Wang, X.; Li, R.; Yang, J.; Gu, D.; Yan, D.; Pu, J.; Chi, B.; Li, J. Effect of YSZ addition on the performance of glass-ceramic seals for intermediate temperature solid oxide fuel cell application. *Int. J. Hydrog. Energy* **2018**, *43*, 8040–8047. [CrossRef]
25. Kumar, V.; Kaur, G.; Pandey, O.P.; Singh, K.; Lu, K. Effect of Thermal Treatment on Chemical Interaction Between Yttrium Borosilicate Glass Sealants and YSZ for Planar Solid Oxide Fuel Cells. *Int. J. Appl. Glas. Sci.* **2014**, *5*, 410–420. [CrossRef]
26. Kaur, G.; Pandey, O.P.; Singh, K. Chemical compatibility between $MgO-SiO_2-B_2O_3-La_2O_3$ glass sealant and low, high temperature electrolytes for solid oxide fuel cell applications. *Int. J. Hydrog. Energy* **2012**, *37*, 17235–17244. [CrossRef]
27. Smeacetto, F.; Salvo, M.; Ferraris, M.; Cho, J.; Boccaccini, A.R. Glass-ceramic seal to join Crofer 22 APU alloy to YSZ ceramic in planar SOFCs. *J. Eur. Ceram. Soc.* **2008**, *28*, 61–68. [CrossRef]
28. Sabato, A.G.; Rost, A.; Schilm, J.; Kusnezoff, M.; Salvo, M.; Chrysanthou, A.; Smeacetto, F. Effect of electric load and dual atmosphere on the properties of an alkali containing diopside-based glass sealant for solid oxide cells. *J. Power Sources* **2019**, *415*, 15–24. [CrossRef]
29. Ferraris, M.; De la Pierre, S.; Sabato, A.G.; Smeacetto, F.; Javed, H.; Walter, C.; Malzbender, J. Torsional shear strength behavior of advanced glass-ceramic sealants for SOFC/SOEC applications. *J. Eur. Ceram. Soc.* **2020**, *40*, 4067–4075. [CrossRef]
30. Li, R.; Liang, X.; Wang, X.; Zeng, W.; Yang, J.; Yan, D.; Pu, J.; Chi, B.; Li, J. Improvement of sealing performance for Al_2O_3 fiber-reinforced compressive seals for intermediate temperature solid oxide fuel cell. *Ceram. Int.* **2019**, *45*, 21953–21959. [CrossRef]
31. Zhang, W.; Wang, X.; Dong, Y.; Yang, J.J.; Pu, J.; Chi, B.; Jian, L. Development of flexible ceramic-glass seals for intermediate temperature planar solid oxide fuel cell. *Int. J. Hydrog. Energy* **2016**, *41*, 6036–6044. [CrossRef]
32. Krainova, D.A.; Saetova, N.S.; Kuzmin, A.V.; Raskovalov, A.A.; Eremin, V.A.; Ananyev, M.V.; Steinberger-Wilckens, R. Non-crystallising glass sealants for SOFC: Effect of Y_2O_3 addition. *Ceram. Int.* **2020**, *46*, 5193–5200. [CrossRef]
33. Krainova, D.A.; Saetova, N.S.; Polyakova, I.G.; Farlenkov, A.S.; Zamyatin, D.A.; Kuzmin, A.V. Behaviour of $54.4SiO_2-13.7Na_2O-1.7K_2O-5.0CaO-12.4MgO-0.6Y_2O_3-11.3Al_2O_3-0.9B_2O_3$ HT-SOFC glass sealant under oxidising and reducing atmospheres. *Ceram. Int.* **2022**, *48*, 6124–6130. [CrossRef]
34. Krainova, D.A.; Saetova, N.S.; Farlenkov, A.S.; Khodimchuk, A.V.; Polyakova, I.G.; Kuzmin, A.V. Long-term stability of SOFC glass sealant under oxidising and reducing atmospheres. *Ceram. Int.* **2021**, *47*, 8973–8979. [CrossRef]
35. Saetova, N.S.; Krainova, D.A.; Kuzmin, A.V.; Raskovalov, A.A.; Zharkinova, S.T.; Porotnikova, N.M.; Farlenkov, A.S.; Moskalenko, N.I.; Ananyev, M.V.; Dyadenko, M.V.; et al. Alumina–silica glass–ceramic sealants for tubular solid oxide fuel cells. *J. Mater. Sci.* **2019**, *54*, 4532–4545. [CrossRef]
36. Kaur, G. *Solid Oxide Fuel Cell Components: Seal Glass for Solid Oxide Fuel Cells*; Springer: Berlin/Heidelberg, Germany, 2016; Volume 58, ISBN 978-3-319-25596-5.
37. Walia, T.; Singh, K. Mixed alkaline earth modifiers effect on thermal, optical and structural properties of $SrO-BaO-SiO_2-B_2O_3-ZrO_2$ glass sealants. *J. Non. Cryst. Solids* **2021**, *564*, 120812. [CrossRef]
38. Kaur, G.; Pandey, O.P.; Singh, K. Effect of modifiers field strength on optical, structural and mechanical properties of lanthanum borosilicate glasses. *J. Non. Cryst. Solids* **2012**, *358*, 2589–2596. [CrossRef]
39. Rodríguez-López, S.; Wei, J.; Laurenti, K.C.; Mathias, I.; Justo, V.M.; Serbena, F.C.; Baudín, C.; Malzbender, J.; Pascual, M.J. Mechanical properties of solid oxide fuel cell glass-ceramic sealants in the system $BaO/SrO-MgO-B_2O_3-SiO_2$. *J. Eur. Ceram. Soc.* **2017**, *37*, 3579–3594. [CrossRef]
40. Lim, E.S.; Kim, B.S.; Lee, J.H.; Kim, J.J. Effect of BaO content on the sintering and physical properties of $BaO-B_2O_3-SiO_2$ glasses. *J. Non. Cryst. Solids* **2006**, *352*, 821–826. [CrossRef]
41. QI, S.; Portnikova, N.M.; Ananyev, M.V.; Kuzmin, A.V.; Eremin, V.A. High-temperature glassy-ceramic sealants $SiO_2—Al_2O_3—BaO—MgO$ and $SiO_2—Al_2O_3—ZrO_2—CaO—Na_2O$ for solid oxide electrochemical devices. *Trans. Nonferrous Met. Soc. China* **2016**, *26*, 2916–2924. [CrossRef]
42. Kaur, N.; Kaur, G.; Kumar, D.; Singh, K. Mechanical and thermal properties of SrO/BaO modified $Y_2O_3-Al_2O_3-B_2O_3-SiO_2$ glasses and their compatibility with solid oxide fuel cell components. *J. Phys. Chem. Solids* **2018**, *118*, 248–254. [CrossRef]
43. Rezazadeh, L.; Hamnabard, Z.; Baghshahi, S.; Golikand, A.N. Adhesion and interfacial interactions of $BaO-SiO_2-B_2O_3$-based glass-ceramic seals and AISI430 interconnect for solid oxide fuel cell applications. *Ionics* **2016**, *22*, 1899–1908. [CrossRef]
44. Navarro-Pardo, F.; Martínez-Barrera, G.; Martínez-Hernández, A.L.; Castaño, V.M.; Rivera-Armenta, J.L.; Medellín-Rodríguez, F.; Velasco-Santos, C. Effects on the thermo-mechanical and crystallinity properties of nylon 6,6 electrospun fibres reinforced with one dimensional (1D) and two dimensional (2D) carbon. *Materials* **2013**, *6*, 3494–3513. [CrossRef] [PubMed]
45. Kerstan, M.; Müller, M.; Rüssel, C. Thermal expansion of $Ba_2ZnSi_2O_7$, $BaZnSiO_4$ and the solid solution series $BaZn_2-xMgxSi_2O_7$ ($0 \leq x \leq 2$) studied by high-temperature X-ray diffraction and dilatometry. *J. Solid State Chem.* **2012**, *188*, 84–91. [CrossRef]
46. Saetova, N.S.; Krainova, D.A.; Kuzmin, A.V. Effect of rare-earth oxides on thermal behavior of alumina-silica glass sealants. *J. Phys. Conf. Ser.* **2021**, *1967*, 012006. [CrossRef]

47. Smiljanić, S.V.; Grujić, S.R.; Tošić, M.B.; Živanović, V.D.; Stojanović, J.N.; Matijašević, S.D.; Nikolić, J.D. Crystallization and sinterability of glass-ceramics in the system La_2O_3-SrO-B_2O_3. *Ceram. Int.* **2014**, *40*, 297–305. [CrossRef]
48. Puig, J.; Ansart, F.; Lenormand, P.; Conradt, R.; Gross-Barsnick, S.M. Development of barium boron aluminosilicate glass sealants using a sol–gel route for solid oxide fuel cell applications. *J. Mater. Sci.* **2016**, *51*, 979–988. [CrossRef]
49. Goel, A.; Reddy, A.A.; Pascual, M.J.; Gremillard, L.; Malchere, A.; Ferreira, J.M.F. Sintering behavior of lanthanide-containing glass-ceramic sealants for solid oxide fuel cells. *J. Mater. Chem.* **2012**, *22*, 10042–10054. [CrossRef]
50. Saetova, N.S.; Shirokova, E.S.; Krainova, D.A.; Chebykin, N.S.; Ananchenko, B.A.; Tolstobrov, I.V.; Belozerov, K.S.; Kuzmin, A.V. The development of 3D technology for the creation of glass sealants for tubular oxide fuel cells. *Int. J. Appl. Glas. Sci.* **2022**, *13*, 684–694. [CrossRef]
51. Ghosh, S.; Kundu, P.; Das Sharma, A.; Basu, R.N.; Maiti, H.S. Microstructure and property evaluation of barium aluminosilicate glass-ceramic sealant for anode-supported solid oxide fuel cell. *J. Eur. Ceram. Soc.* **2008**, *28*, 69–76. [CrossRef]
52. Peng, L.; Zhu, Q.S. Thermal cycle stability of BaO-B_2O_3-SiO_2 sealing glass. *J. Power Sources* **2009**, *194*, 880–885. [CrossRef]
53. Da Silva, M.J.; Bartolomé, J.F.; De Aza, A.H.; Mello-Castanho, S. Glass ceramic sealants belonging to BAS (BaO-Al_2O_3-SiO_2) ternary system modified with B2O3 addition: A different approach to access the SOFC seal issue. *J. Eur. Ceram. Soc.* **2016**, *36*, 631–644. [CrossRef]
54. Li, X.; Yazhenskikh, E.; Groß-Barsnick, S.M.; Baumann, S.; Behr, P.; Deibert, W.; Koppitz, T.; Müller, M.; Meulenberg, W.A.; Natour, G. Crystallization behavior of BaO–CaO–SiO_2–B_2O_3 glass sealant and adjusting its thermal properties for oxygen transport membrane joining application. *J. Eur. Ceram. Soc.* **2023**, *43*, 2541–2552. [CrossRef]
55. Craievich, A.F.; Zanotto, E.E.; James, P.F. Kinetics of sub-liquidus phase separation in silicate and borate glasses. A review. *Bull. Mineral.* **1983**, *106*, 169–184. [CrossRef]
56. Abdel-Hameed, S.A.M.; Abo-Naf, S.M.; Hamdy, Y.M. The effect of heat treatment on photoluminescence and magnetic properties of new yellow phosphor based on sanbornite ($BaSi_2O_5$) glass ceramic doped with Gd^{3+} and Mn^{2+}. *J. Non. Cryst. Solids* **2019**, *517*, 106–113. [CrossRef]
57. Saxena, S.K.; Chatterjee, N.; Fei, Y.; Shen, G. *Thermodynamic Data on Oxides and Silicates: An Assessed Data Set Based on Thermochemistry and High Pressure Phase Equilibrium*; Springer: Berlin/Heidelberg, Germany, 1993; ISBN 3642783325.
58. Decterov, S.A.; Jung, I.-H.; Pelton, A.D. Thermodynamic Modeling of the FeO-Fe_2O_3-MgO-SiO_2 System. *J. Am. Ceram. Soc.* **2002**, *85*, 2903–2910. [CrossRef]
59. Jung, I.H.; Decterov, S.A.; Pelton, A.D.; Kim, H.M.; Kang, Y.B. Thermodynamic evaluation and modeling of the Fe-Co-O system. *Acta. Mater.* **2004**, *52*, 507–519. [CrossRef]
60. Kaur, G.; Singh, K.; Pandey, O.P. Investigations on Interfacial Interaction of Glass Sealants with Electrolytes and Interconnect for Solid Oxide Fuel Cells. Ph.D. Thesis, Thapar University, Punjab, India, 2012.
61. Brochu, M.; Gauntt, B.D.; Shah, R.; Miyake, G.; Loehman, R.E. Comparison between barium and strontium-glass composites for sealing SOFCs. *J. Eur. Ceram. Soc.* **2006**, *26*, 3307–3313. [CrossRef]
62. Chukhlantsev, V.G.; Yu, M. Galkin Study of the BaO-ZrO_2-SiO_2 System at Subsolidus temperatures. *Dokl. Akad. Nauk. SSSR* **1968**, *168–169*, 128.
63. Li, B.; Tang, B.; Xu, M. Influences of CaO on Crystallization, Microstructures, and Properties of BaO-Al_2O_3-B_2O_3-SiO_2 Glass–Ceramics. *J. Electron. Mater.* **2015**, *44*, 3849–3854. [CrossRef]
64. Li, B.; Xu, M.; Tang, B. Effects of ZnO on crystallization, microstructures and properties of BaO–Al_2O_3–B_2O_3–SiO_2 glass–ceramics. *J. Mater. Sci. Mater. Electron.* **2016**, *27*, 70–76. [CrossRef]
65. Kermani, P.S.; Ghatee, M.; Sirr Irvine, J.T. Characterization of a barium-calcium-aluminosilicate glass/fiber glass composite seal for intermediate temperature solid oxide fuel cells. *Bol. La Soc. Esp. Ceram. Y Vidr.* **2022**, in press. [CrossRef]
66. Lu, X.; Deng, L.; Kerisit, S.; Du, J. Structural role of ZrO_2 and its impact on properties of boroaluminosilicate nuclear waste glasses. *Npj Mater. Degrad.* **2018**, *2*, 19. [CrossRef]
67. Khan, S.; Kaur, G.; Singh, K. Effect of ZrO_2 on dielectric, optical and structural properties of yttrium calcium borosilicate glasses. *Ceram. Int.* **2017**, *43*, 722–727. [CrossRef]
68. Chen, S.; Yu, Z.; Zhang, Q.; Wang, J.; Zhang, T.; Wang, J. Reducing the interfacial reaction between borosilicate sealant and yttria-stabilized zirconia electrolyte by addition of HfO_2. *J. Eur. Ceram. Soc.* **2015**, *35*, 2–6. [CrossRef]

Disclaimer/Publisher's Note: The statements, opinions and data contained in all publications are solely those of the individual author(s) and contributor(s) and not of MDPI and/or the editor(s). MDPI and/or the editor(s) disclaim responsibility for any injury to people or property resulting from any ideas, methods, instructions or products referred to in the content.

Article

Phase Transformations upon Formation of Transparent Lithium Alumosilicate Glass-Ceramics Nucleated by Yttrium Niobates

Olga Dymshits [1,*], Anastasia Bachina [2], Irina Alekseeva [1], Valery Golubkov [3], Marina Tsenter [1], Svetlana Zapalova [1], Kirill Bogdanov [4], Dmitry Danilovich [5] and Alexander Zhilin [6]

[1] S.I. Vavilov State Optical Institute, 36 Babushkina St., 192171 St. Petersburg, Russia; vgolub19@gmail.com (I.A.); myzenter@gmail.com (M.T.); zenii99@yandex.ru (S.Z.)
[2] Ioffe Institute, Russian Academy of Sciences, 194021 St. Petersburg, Russia; a.k.bachina@yandex.ru
[3] Institute of Silicate Chemistry of Russian Academy of Sciences, Adm. Makarova emb. 2, 199034 St. Petersburg, Russia; nitiom@goi.ru
[4] Center of Information Optical Technologies, ITMO University, Kronverkskiy pr. 49, 197101 St. Petersburg, Russia; kirw.bog@gmail.com
[5] Department of Chemical Technology of Refractory Nonmetallic and Silicate Materials, Faculty of Chemistry of Substances and Materials, St. Petersburg State Technological Institute (Technical University), Moskovski Ave. 26, 190013 St. Petersburg, Russia; danilovich@technolog.edu.ru
[6] D.V. Efremov Institute of Electrophysical Apparatus, Metallostroy, Doroga na Metallostroy, 3 Bld., 196641 St. Petersburg, Russia; zhilin1311@yandex.ru
* Correspondence: vodym1959@gmail.com; Tel.: +7-921-8640217

Abstract: Phase transformations in the lithium aluminosilicate glass nucleated by a mixture of yttrium and niobium oxides and doped with cobalt ions were studied for the development of multifunctional transparent glass-ceramics. Initial glass and glass-ceramics obtained by isothermal heat-treatments at 700–900 °C contain $YNbO_4$ nanocrystals with the distorted tetragonal structure. In samples heated at 1000 °C and above, the monoclinic features are observed. High-temperature X-ray diffraction technique clarifies the mechanism of the monoclinic yttrium orthoniobate formation, which occurs not upon high-temperature heat-treatments above 900 °C but at cooling the glass-ceramics after such heat-treatments, when $YNbO_4$ nanocrystals with tetragonal structure undergo the second-order transformation at ~550 °C. Lithium aluminosilicate solid solutions (ss) with β-quartz structure are the main crystalline phase of glass-ceramics prepared in the temperature range of 800–1000 °C. These structural transformations are confirmed by Raman spectroscopy and illustrated by SEM study. The absorption spectrum of the material changes only with crystallization of the β-quartz ss due to entering the Co^{2+} ions into this phase mainly in octahedral coordination, substituting for Li^+ ions. At the crystallization temperature of 1000 °C, the Co^{2+} coordination in the β-quartz solid solutions changes to tetrahedral one. Transparent glass-ceramics have a thermal expansion coefficient of about 10×10^{-7} K^{-1}.

Keywords: yttrium niobate; β-quartz solid solution; glass-ceramics; nanocrystals; small-angle X-ray scattering; in situ high-temperature X-ray diffraction; scanning electron microscopy; Raman spectroscopy; optical spectroscopy

1. Introduction

Low thermal expansion transparent glass-ceramics of the lithium aluminosilicate (LAS) system form commercially the most important glass-ceramic family [1,2]. They are composed of the main crystalline phase of lithium aluminosilicate solid solutions (ss) with β-quartz structure and have found their applications as cooktop plates, woodstove windows, fireplaces, as cooking ware and fire protection doors or windows, large telescope mirror blanks, ring laser gyroscopes, liquid crystal displays, and optical components [1,2]. A combination of unique thermal-mechanical properties of LAS glass-ceramics and their transparency with optical properties of rare-earth ions in oxide nanocrystals could allow for the development of new multifunctional materials. However, ionic radii of rare-earth ions are larger than those of ions that constitute the β-quartz ss, and there exist no cation sites for trivalent rare-earth ions in the LAS crystals [3]; β-quartz ss appeared to be a poor host for fluorescent cations [4].

It should be noted that LAS glass-ceramics are multiphase materials that contain the crystalline phases of not only the β-quartz ss, but also of the nucleating agent and it is the nucleating agent that can become a promising host for trivalent rare-earth ions. In our previous studies [5–7], we developed new transparent LAS glass-ceramics containing rare-earth orthoniobates, such as (Er,Yb)NbO$_4$, YbNbO$_4$ [6], and (Eu^{3+},Yb^{3+}):YNbO$_4$ [7], bearing a bifunctional role of nucleating agents and luminescent compounds. It was found that nanocrystals of rare-earth orthoniobates with disordered fluorite structures precipitate from the initial X-ray amorphous glass upon heat-treatment; at elevated temperatures, structural transformation to a tetragonal phase takes place. Glass-ceramics prepared by heat-treatments at above 950 °C additionally contain crystals of rare-earth orthoniobates with a monoclinic structure. The spectral-luminescent properties of rare-earth ions are directly linked to the structure of corresponding rare-earth orthoniobates [5–7]. However, it was not clear up to now if the monoclinic rare-earth orthoniobates were formed during the high-temperature heat-treatments of initial glasses or upon cooling down the glass-ceramics prepared by these heat-treatments. Therefore, the aim of the present study is to clarify the sequence and the mechanism of phase transformations in transparent glass-ceramics containing rare-earth and niobium oxides by means of in situ high-temperature X-ray diffraction and differential thermal analysis.

In this study, yttrium was taken as a representative rare-earth ion, and yttrium orthoniobate was chosen as a model rare-earth orthoniobate. Yttrium orthoniobate is a promising optical host for rare-earth ions because yttrium ions are easily replaced in any proportion by other rare-earth ions with similar ionic radii [7–15]. It should be noted that yttrium orthoniobate itself is distinguished by the combination of promising luminescent, chemical, and mechanical properties. It is a self-activated X-ray phosphor [16] and widely used in X-ray medical techniques. We believe the regularities of phase transformations found by the example of yttrium orthoniobates crystallized in LAS glass-ceramics are the general regularities that can be extended to other rare earth orthoniobates crystallized in LAS glass-ceramics.

In nature, yttrium orthoniobate, YNbO$_4$, is known as a mineral fergusonite [17,18]. There are three crystalline forms of synthetic yttrium orthoniobate, monoclinic (M-phase, M-fergusonite) at room temperature, tetragonal (T-phase) with the scheelite (CaWO$_4$) structure at higher temperatures [19], and a high-temperature cubic phase [20]. The structural transformation between the monoclinic and tetragonal forms of YNbO$_4$ is reversible and proceeds by a gradual change in symmetry in the approximate range of 500 to 800 °C; the T-phase cannot be preserved at room temperature [19]. These phase transformations were studied by K. Jurkschat et al. [20] and considered to be displacive ferroelastic transformations of a second order, leading to the formation of ferroelastic twinning domains during the T–M phase transition. Heating of YNbO$_4$ crystals close to the melting temperature shows that a high-temperature cubic phase could be detected, which consists of a solid solution of YNbO$_4$ and Y$_3$NbO$_7$ and possibly small amounts of Nb$_2$O$_3$ [20].

While heating an amorphous material prepared by the simultaneous hydrolysis of yttrium and niobium alkoxides [21] or by the sol-gel method [22], another form of YNO$_4$, the T' one, was crystallized. This phase has been shown [22] to have a distorted tetragonal structure similar to that found in tetragonal ZrO$_2$ with pseudo-fluorite lattice. The T' to M phase transformation does not occur, and the T'-phase can be obtained at room temperature [21]. O. Yamaguchi et al. [21] suggested that the distortion of the tetragonal structure is responsible for stabilization of the T'-phase at room temperature. The heating behavior of this phase is very similar to that of the metamict mineral; it crystallizes in a tetragonal modification when heated for prolonged periods of time at 400 to 800 °C, and when heated to 1000 °C and cooled it is monoclinic [21].

Here, for the first time we report the preparation of transparent glass-ceramics of the LAS system containing yttrium orthoniobates nanocrystals by the melt-quenching method with the aim to clarify the sequence and the mechanism of phase transformations of yttrium orthoniobate nanocrystals in this system by means of in situ high-temperature X-ray diffraction and differential thermal analysis.

2. Materials and Methods

2.1. Materials Preparation

For this study, we have chosen the glass of the composition 18 Li_2O, 27 Al_2O_3, 55 SiO_2 (mol%) [23], 5 mol% Y_2O_3, 5 mol% Nb_2O_5, and 0.1 mol% CoO were added above 100% as nucleators and colour dopant, respectively. The reagent-grade raw materials were supplied by Nevareactive, Saint Petersburg, Russia. The batch weight was 300 g. The glass was melted in air in a home-made crucible made of quartz ceramics at 1560 °C for 4 h with stirring, and then poured out onto a metal plate and annealed at 620 °C for 1 h. Then, the furnace was switched off and the glass was cooled down with the annealing furnace. The transparent violet-blue-colored initial glass was heat-treated in isothermal conditions from 700 to 1350 °C for 6 h by the two-stage heat-treatments with the first hold at 700 °C for 6 h. The heating rate was 5 °C per minute.

2.2. Characterization

2.2.1. Thermal Analysis

Differential thermal analysis (DTA) was carried out using a simultaneous thermal analyzer, Netzsch STA 449F3 Jupiter, in platinum crucibles. The measurements were taken upon heating from 30 to 1200 °C and upon cooling from 1200 to 300 °C under the argon flow (30 mL/min). Two sets of measurements were conducted, employing heating and cooling rates of 10 and 30 °C/min, respectively. The sample weight was 15 mg.

2.2.2. Powder X-ray Diffraction (PXRD)

PXRD measurements were performed using a Shamadzu XRD-6000 diffractometer with CuKα radiation (λ = 1.5406 Å). The phase composition was analyzed by matching the recorded PXRD patterns with the Inorganic Crystal Structure Database (ICSD). The Rietveld refinement was performed using PDWin 4.0 software (Burevestnik, St. Petersburg, Russia). The mean crystal sizes were estimated from broadening of X-ray peaks according to Scherrer's equation:

$$D = K\lambda / \Delta(2\theta)\cos\theta, \qquad (1)$$

where λ is the wavelength of X-ray radiation, θ is the diffraction angle, Δ(2θ) is the width of peak at half of its maximum, and K is the constant assumed to be 1 [24]. The error in the mean crystal size estimation is about 5%. The size of T'- and T-forms of $YNbO_4$ nanocrystals was estimated using the peak with the Miller's indices (hkl) of (112). The size of the β-quartz ss was estimated using the peak with the (hkl) indices of (121).

The a and c lattice parameters of the T'- and T-forms of $YNbO_4$ crystals were estimated from the positions of diffraction peaks with the Miller's indices (hkl) of (112) and (004), which ensured accuracy (±0.003 Å), according to the equation:

$$1/d_{hkl}^2 = (h^2 + k^2)/a^2 + l^2/c^2. \qquad (2)$$

High-temperature powder X-ray diffraction (HT-PXRD) patterns were recorded with a Shimadzu XRD-7000 diffractometer with CuKα radiation (λ = 1.5406 Å) equipped with an Anton Paar HTK-1200 furnace attachment. Powdered samples were heat-treated from room temperature up to 1100 °C with a step of 100 °C, a heating/cooling rate of 30 °C per minute, and isothermal holdings of 15 min at each step. PXRD patterns were collected in-situ after each isothermal holding upon heating and cooling. The measurements were taken for the Bragg angle (2θ) range from 20 to 40° (i.e., over the range of the strongest peaks of possible crystalline phases in the material studied) with a step of 0.02° and a scan rate of 2°/min.

2.2.3. Small Angle X-ray Scattering

Plane-parallel polished samples with thickness of 0.2 mm were studied by small angle X-ray scattering (SAXS) method. The SAXS intensity $I(\phi)$ was measured with a home-made instrument in the range of scattering angles ϕ from 6 to 450 arc min. CuKα radiation (λ = 1.5406 Å) was used with an "infinitely" high primary beam ("infinitely" high slit).

SAXS curves of glass-ceramics and phase-separated glasses often exhibit maximum, which appears due to the regularity in a distribution of scattering regions in the glass volume. In the presence of spatial ordering of the scattering regions, at which the interference maximum is observed, the value of the angle ϕ_m corresponding to the maximum on the $I(\phi)$ curve ϕ_m and the mean distance between centers of regularly distributed particle (L) are related as

$$L \cong (2.85 \div 2.97) \cdot 10^3 \, \phi_m^{-1}, \quad (3)$$

where L (in Å) is a distance between centers of regularly distributed particle [25]. The position of the maximum ϕ_m on the ϕ dependence of $\phi \cdot I(\phi)$ corresponds to the inverse radius of the scattering regions. For the monodisperse system of spherical particles with radius R, in condition of validity of Guineir equation [26],

$$R = 1328/\phi_m. \quad (4)$$

2.2.4. Scanning Electron Microscopy (SEM) and Energy Dispersive X-ray (EDX)-Based Element Analysis

The morphology of glass-ceramics was characterized by scanning electron microscopy (SEM) using Tescan Vega 3 SBH microscope. For the study, the surface of the polished samples was preliminary etched in a hydrofluoric acid for about 10 s. Particle size was calculated using ImageJ software. The same Tescan Vega 3 SBH microscope equipped with an Oxford INCA 200 energy-dispersive detector was employed for energy dispersive X-ray (EDX)-based element analysis.

2.2.5. Raman Spectroscopy

Raman spectra were recorded on plane-parallel polished samples with a thickness of ~1 mm in backscattering geometry by using an InVia (Renishaw, Wotton-under-Edge, UK) Micro-Raman spectrometer equipped with the CCD camera cooled up to −70 °C. Ar$^+$ laser line of 514 nm was employed as an excitation source. Leica 50 × (NA = 0.75) objective was used for illuminating the sample; the scattered light was collected by the same objective. Edge filter was placed before the spectrograph entrance slit. A spatial resolution of 2 cm^{-1} was obtained. Acquisition time was 60 s.

2.2.6. Absorption Spectroscopy

Absorption spectra were measured on a Shimadzu UV-3600 spectrophotometer in the spectral range from 200 to 3300 nm on the same plane-parallel polished samples with a thickness of ~1 mm that were used for the Raman spectra recording. The wavelength step of measurements was 0.5 nm.

2.2.7. The Linear Coefficient of Thermal Expansion

The linear coefficient of thermal expansion (CTE) was measured with the Linseis L75 VS 1000 dilatometer using samples with the length of 25–35 mm and cross-section of 3 × 3 mm. The measurements were taken upon heating from 30 to 500 °C at a heating rate of 5 K/min.

3. Results

The images of polished samples of transparent initial glass and glass-ceramics with the thickness of 1 mm and of the opaque sample prepared by the heat-treatment at 700 °C and at 1350 °C for 6 h at each stage are shown in Figure 1.

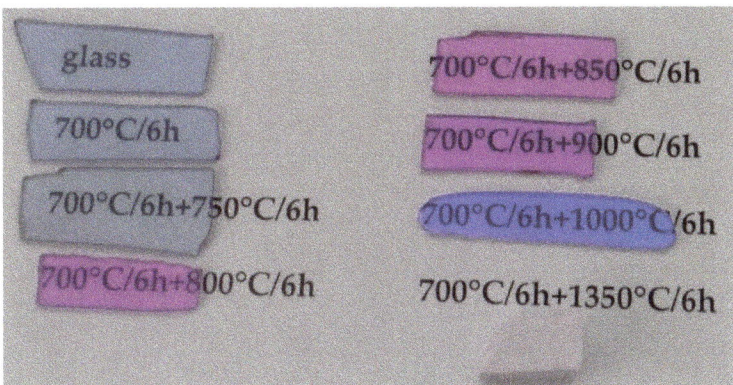

Figure 1. Images of samples of the initial and heat-treated glasses. The thickness of polished transparent samples is 1 mm. The values indicated in the figure denote heat-treatment schedules.

3.1. DTA and XRD Studies

3.1.1. DTA Study

The first set of DTA measurements was conducted using conventional heating and cooling rates of 10 °C/min. Several thermal effects were observed: a glass transition at 697 °C, followed by three exothermic peaks with onset temperatures at 782 °C (broad and non-intense), 828 °C (sharp and intense), and 913 °C (very weak). No significant effects were observed during the cooling process. With the aim to reveal the possible thermal effect of the tetragonal–monoclinic phase transition upon cooling, we increased the heating and cooling rate, and the second set of DTA measurements was performed at a heating and cooling rate of 30 °C/min. The increased heating and cooling rate resulted in a shift of the thermal effects towards higher temperatures, and two additional weak features were revealed on the DTA curves. Figure 2 displays the data obtained from the second set of DTA measurements. Consequently, the DTA curve of the initial glass exhibited a glass transition temperature (T_g) at 708 °C and four exothermic peaks during heating (Figure 2). The first broad exothermic peak occurred at T_{on1} = 812 °C, where T_{on1} is a crystallization onset temperature, indicating the onset of crystallization of the nucleator in the form of the T'-phase. This peak was followed by a sharp and intense peak at T_{on2} = 861 °C, corresponding to the crystallization of β-quartz ss. Two small, broad peaks were observed at T_{on3} = 935 °C and T_{on4} = 1097 °C. The peak at T_{on3} = 935 °C was attributed to the T'-phase to T-phase transformation, while the small peak at T_{on4} = 1097 °C was associated with the formation of β-spodumene ss. The DTA curve collected on cooling showed an inflection point at temperature of 547 °C. This observation suggests the existence of a weak thermal effect, which may be attributed to the phase transition from the T-phase to the M-phase, which is known to be a continuous second-order transition [27]. To further support this suggestion, HT-PXRD investigations were conducted and the results are described in detail below.

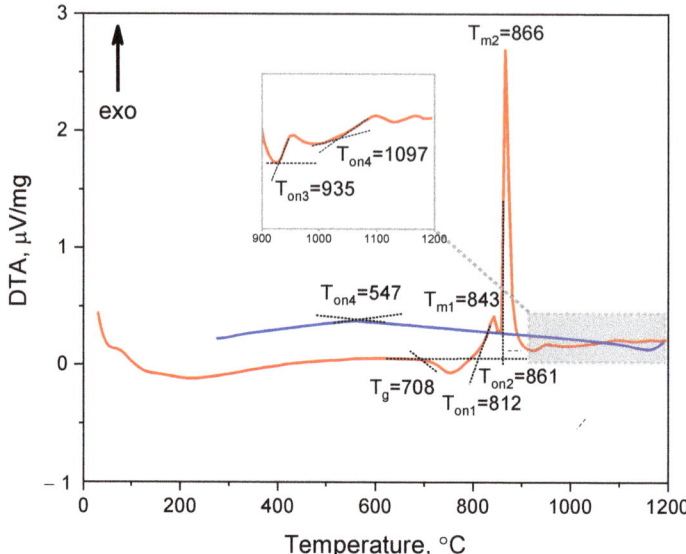

Figure 2. The DTA curve of the initial glass recorded in heating (red curve) and cooling (blue curve) modes. The heating and cooling rates are 30 °C/min. The values indicated in the figure denote the temperature, in °C.

3.1.2. Isothermal Heat-Treatments

The XRD patterns for the initial and heat-treated glasses are presented in Figure 3. Initial glass, as well as the glass heat-treated at 700 °C for 6 h (Figure 3a) contains the crystals of yttrium orthoniobate, $YNbO_4$, with distorted tetragonal structure (T'-phase) 11 nm in size (Table 1). The addition of the second heat-treatment hold at 750 °C for 6 h results in increasing fraction of the T'-phase with the same crystal size. After heat-treatments with the second hold at 800–1000 °C, volume crystallization of lithium aluminosilicate solid solutions (ss) with β-quartz structure (β-quartz ss) was observed. Thus, nanocrystals of yttrium orthoniobates serve as the heterogeneous nucleation sites for the crystallization of β-quartz ss. The crystallinity fractions and crystal sizes increase with an increase in the heat-treatment temperature. After heat-treatment at 900 °C for 6 h, the size of T'-$YNbO_4$ crystals is 12 nm, the size of β-quartz ss is about 42 nm. XRD pattern of the sample heat-treated at the second stage at 1000 °C for 6 h shows co-precipitation of tetragonal and monoclinic nanocrystals (M-phase) of $YNbO_4$ together with β-quartz ss (Figure 3a). The lattice parameters of the T'-phase are consistent with previously reported values of a = 5.164 Å, c = 10.864 Å [21] and show no significant change until the temperature of the second hold reaches 900 and 1000 °C, see Table 1. At the second stage at 1000 °C, the appearance of the M-phase is accompanied by a shift towards lower Bragg angles of the peaks in the XRD pattern, corresponding to the tetragonal yttrium niobate (Figure 3a). This observation is reflected in the lattice parameters, which become larger and closer to the values a = 5.21 Å and c = 11.05 Å of lattice parameters of the high-temperature T-phase [21,28].

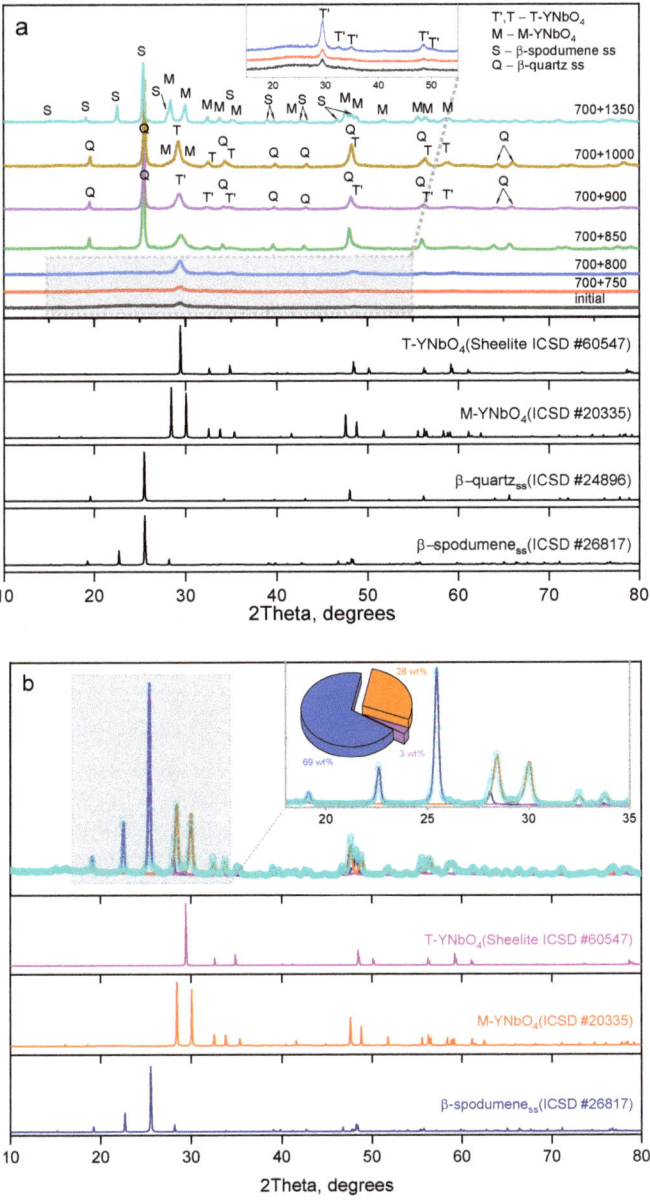

Figure 3. XRD patterns of the initial and heat-treated glasses: (**a**) initial glass and glasses subjected to two-stage heat treatments at indicated temperatures, °C. The appropriate letters are used to indicate the primary peaks of the crystalline phases. (**b**) results of the phase analysis of the glass heat-treated at temperatures of 700 °C + 1350 °C using Rietveld method (R_{wp} = 8.35%). The inserts display the magnified Bragg angle ranges containing the most intense peaks of all crystalline phases for improved visibility. The pie plot represents the quantified phase composition (mass%), where blue corresponds to β-spodumene ss, orange to the M-phase, and light violet to the T-phase. The duration of each hold is 6 h. The diffraction patterns are shifted along the vertical axis for the convenience of observation.

Table 1. Characteristics of crystalline phases of yttrium niobate and β-quartz ss in LAS glass-ceramics prepared by different heat-treatments.

Heat-Treatment Schedule, °C/h	Yttrium Niobate			β-Quartz ss
	Crystal Size, nm	Lattice Parameters, Å		Crystal Size, nm
		a, ±0.003	c, ±0.003	
Initial glass	11.0 ± 0.3	5.120	11.03	-
700/6	10.0 ± 0.3	5.121	11.08	-
700/6 + 750/6	10.5 ± 0.3	5.128	11.06	-
700/6 + 800/6	10.0 ± 0.3	5.120	11.03	46.0 ± 1.5
700/6 + 900/6	12.0 ± 0.3	5.157	11.04	42.0 ± 1.0
700/6 + 1000/6	15.5 ± 0.4	5.158	11.01	40.5 ± 1.0

The sample heat-treated at 700 °C and 1350 °C for 6 h at each stage contains monoclinic $YNbO_4$ crystals with the admixture of the T-phase and crystals of β-spodumene ss (Figure 3b). To refine its structure, a Rietveld refinement was performed for this sample, and the results are presented in Figure 3b. The initial models for structure refinement of the monoclinic phase of $YNbO_4$, the tetragonal phase of $YNbO_4$, and β-spodumene ss were based on ICSD records #20335, #60547, and #26817, respectively. According to the phase analysis, the composition is as follows: 69 ± 1 mass% of lithium aluminosilicate with a structure of β-spodumene ss, 28 ± 1 mass% yttrium niobate with a monoclinic structure (M-$YNbO_4$), and a trace amount of 3 ± 1 mass% yttrium niobate with a tetragonal structure (T-$YNbO_4$). The refined lattice parameters for the β-spodumene ss phase (space group: $P4_32_12$) are a = 7.5357(9) Å and c = 9.1689(15) Å, while for the M-$YNbO_4$ phase (space group: $C12/c1$), the lattice parameters are a = 7.0521(10) Å, b = 10.9788(17) Å, c = 5.3004(8) Å, and β = 134.180(5)°. The refined lattice parameters for the β-spodumene ss and the M-phase are consistent with previous reports [29,30], respectively.

3.1.3. In Situ High-Temperature XRD Study

According to Figure 4a, above 700 °C, nanocrystalls of T'-$YNbO_4$ continue to grow from the initial glass containing traces of the T'-phase of yttrium niobate. Crystallization of β-quartz ss begins at 800 °C. The temperature raise to 1100 °C leads to the transformation of the distorted tetragonal T'-$YNbO_4$ to the high-temperature sheelite-like tetragonal structure. Thus, XRD patterns collected in situ upon heating demonstrate no evidence of low-temperature monoclinic phase (M-$YNbO_4$) and argue for sequential transformations of the yttrium niobate phase from the metastable distorted tetragonal T'-phase to the thermodynamically stable at high temperatures sheelite-like one. Nanocrystals of M-$YNbO_4$ are formed from the high-temperature tetragonal phase with sheelite-like structure upon cooling in the temperature range between 600 °C and room temperature, see Figure 4b.

3.2. Raman Spectroscopy Study

The Raman spectrum of the initial glass demonstrates the broad band spanning from ~200 to 350 cm^{-1}, the band at ~480 nm, an intense band with the maximum at 806 cm^{-1} and a shoulder in the range of about 880–1000 cm^{-1}, see Figure 5. The spectrum of the glass heat-treated at 700 °C for 6 h resembles that of the initial glass, which is explained by their similar phase compositions, see Figure 3a. The Raman spectrum drastically changes after the heat-treatment at 700 + 1000 °C for 6 h at each stage. There are bands at 211, 235, 327, 335, 422, 438, 482, 661, and 693 cm^{-1}, and the most pronounced band is at 806 cm^{-1} (Figure 5).

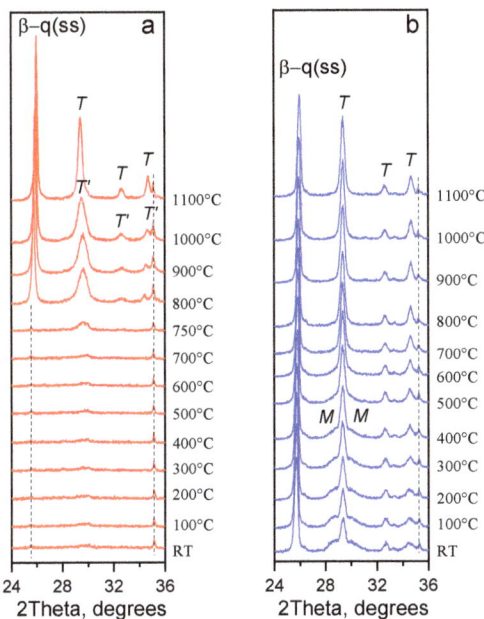

Figure 4. In-situ PXRD patterns collected during (**a**) the heating of the initial glass and (**b**) the cooling of the fabricated glass-ceramic. The assigned peaks correspond to the following phases: T'—distorted, disordered tetragonal phase of $YNbO_4$; T—high-temperature ordered tetragonal phase of $YNbO_4$ with sheelite structure; M—M-fergusonite-like monoclinic phase of $YNbO_4$; β-q(ss)—lithium aluminosilicate solid solution with β-quartz structure. The dashed lines show the lack of peaks shift of corundum from the sample holder. The diffraction patterns are shifted along the vertical axis for the convenience of observation.

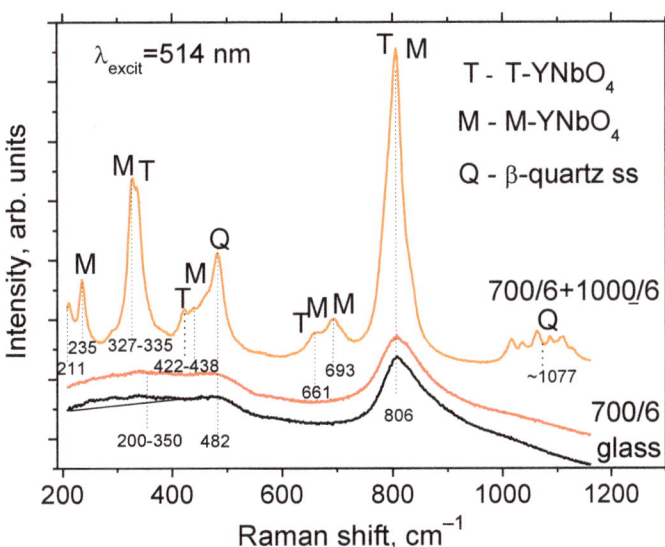

Figure 5. Raman spectra of the initial and heat-treated glasses. The excitation wavelength is 514 nm. The values indicated in the figure denote heat-treatment temperatures and time, °C/h. The dashed lines show positions of the maxima, cm^{-1}. The spectra are shifted for the convenience of observation.

In the Raman spectrum of the initial glass (Figure 5), the broad band of about 480 cm^{-1} and a shoulder at 880 to 900 cm^{-1} are due to vibrations of the aluminosilicate glass network with a large number of non-bridging oxygens [31] and of [NbO$_4$] and [NbO$_6$] polyhedrons in the glass structure [32], respectively. The broad band at ~1000 cm^{-1} corresponds to vibrations of [SiO$_4$] tetrahedrons in the aluminosilicate glass network. A very weak broad band at about 200–350 cm^{-1}, as well as the intense band at 806 cm^{-1}, reveal the beginning of formation of distorted orthoniobate T'-YNbO$_4$ crystals with disordered fluorite-type structure [6,7,33]. The Raman spectrum of the glass-ceramic obtained by the heat-treatment at 1000 °C for 6 h gives the evidence of the absence of T'-phase (the disappearance of the broad band at about 200–350 cm^{-1}). A number of distinct bands appear in the Raman spectrum of the sample. The bands at 482 and 1077 cm^{-1} reveal crystallization of the β-quartz ss [34]. The complex shape of the broad band with the maximum at ~1077 cm^{-1} is due to luminescence of Er^{3+} ions, the unwanted impurity in the reagent of yttrium oxide [7]. The bands at ~330, ~420, ~660, and 806 cm^{-1} belong to vibrations of [NbO$_4$] tetrahedrons in the T-YNbO$_4$ crystals [35]. Since the frequencies of the main Raman bands of the T- and M-YNbO$_4$ crystals are very close [35], unambiguous judgment about the appearance of the M-phase is possible by the appearance in the Raman spectrum of two intense bands at 235 and 690 cm^{-1} and two weak bands at ~211 and ~435 cm^{-1}. Thus, according to Raman spectroscopy findings in the sample heat-treated at 700 +1000 °C for 6 h, the tetragonal and monoclinic crystals of YNbO$_4$ coexist, which is in accordance with the XRD findings, see Figure 3a.

3.3. Small Angle X-ray Scattering Study

Figure 6 shows the angular dependences of the SAXS intensity for the initial glass and glasses heat-treated at 700 °C and at 700 + 1000 °C. Due to high electron density of the inhomogeneous regions containing nanosized YNbO$_4$ crystals, the small-angle X-ray scattering is predominantly determined by scattering by these regions.

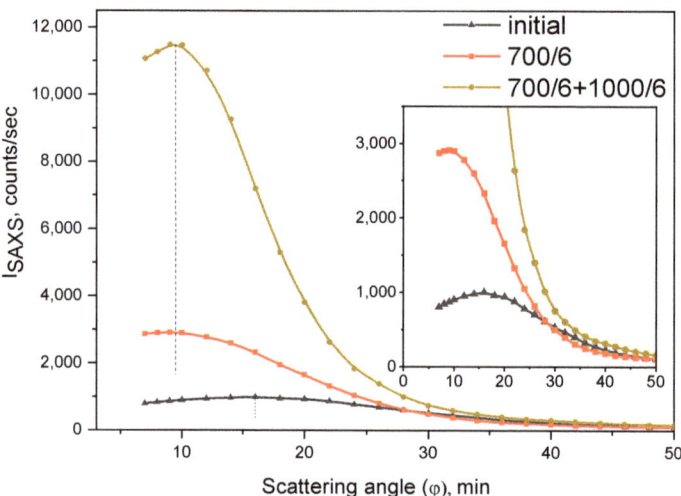

Figure 6. Small angle X-ray scattering of the initial and heat-treated glasses. The values indicated in the figure denote heat-treatment temperatures and time, °C/h. The dashed line shows the position of the maximum.

The initial glass demonstrates a well-developed inhomogeneous structure. It contains inhomogeneity regions with radii of ~50 Å as the result of phase separation and crystallization during the melt casting and cooling. The SAXS intensity increases with heat-treatment due to continuous crystallization of YNbO$_4$ nanocrystals.

All SAXS curves exhibit a maximum, which indicates interparticle interference and, accordingly, ensures an order in the distribution of regions of inhomogeneity. It confirms the role of yttrium niobate as a nucleating agent. It can be assumed that interference effects affect the intensity of light scattering and transparency.

The maximum on the SAXS curve from the initial glass presented in Figure 6 is observed at a scattering angle of about 16′. So the average distance between scattering regions estimated by the Equation (3) is about 180 Å. As can be seen from Figure 6, low-temperature heat-treatment at 700 °C for 6 h leads to a shift of the maximum to smaller angles of ~10′ and, accordingly, to an increase in the distance between the inhomogeneous regions to ~290 Å. After the heat-treatment at 700°C + 1000 °C for 6 h, the position of the maximum on the SAXS curve practically does not change, that is, the geometry of the structure and the distribution of regions of inhomogeneity do not change. The radii of scattering regions increase from ~50 Å for the initial glass to ~80 Å for the glass heat-treated at 700 °C for 6 h and to ~100 Å for the glass-ceramics prepared by the heat-treatment at 1000 °C at the second stage. The main reasons for the increase in the intensity of SAXS during high-temperature heat-treatments are structural transformations within the inhomogeneous regions resulting in the crystallization of the $YNbO_4$ phase.

3.4. SEM-EDX Study

Figure 7 illustrates the evolution of the initial glass morphology as a function of the heat-treatment temperature. Analysis of the SEM data of the initial glass sample revealed the presence of regions of inhomogeneity in the form of spherical particles characterized by a narrow size distribution. The average size of these particles was determined to be approximately 24 nm. Based on our previous studies [6,7] and the present data of the SAXS analysis, we suggest that nanocrystals of $YNbO_4$ are formed within these regions of inhomogeneity. Upon subjecting the samples to two-stage heat treatments with increasing temperature, an increase in the number of these regions was observed. Their size distributions exhibited a broader shape, and the maxima of the distributions shifted towards higher values.

The average size of the regions of inhomogeneity in the sample heat-treated at the second stage at 850 °C was approximately 35 nm, while in the sample heat-treated at temperature of 1000 °C, the average size was approximately 37 nm. The morphology of the sample, which underwent a two-stage heat-treatment at 700 °C + 1350 °C, exhibited a drastic change compared to the morphology of the initial glass and the other glass-ceramics. In Figure 7d, two distinct types of particles were observed. The particles of the first type were almost spherical; with a bimodal size distribution (the smaller particles had the diameter of ~0.4 μm and the larger ones of ~0.8 μm) with the average diameter of approximately 0.72 μm. The particles of the second type had a rod-like shape with rectangular section. The length of these rod-like particles ranged from 0.2 to 6.5 μm, while the width varied from 73 to 266 nm. Based on Refs. [27,36,37] and our XRD findings, we believe that the almost spherical particles correspond to yttrium orthoniobate crystals with a monoclinic structure. Some of them have distinct facets, while the other show the melted facets, see Figure 7d. We attribute the rod-like particles to the crystalline phase of β-spodumene ss.

The microstructure and elemental composition of glass-ceramic prepared by the heat-treatment at 700 °C + 1350 °C were evaluated by SEM coupled with EDX analysis. The results are shown in Figure 8a–c. It should be mentioned that the lithium content cannot be determined by the EDX method, and the content of all other elements was normalized to 100%.

Figure 7. SEM images of the samples of (**a**) the initial glass, (**b**–**d**) heat-treated at (**b**) 700 °C + 850 °C, (**c**) 700 °C + 1000 °C, and (**d**) 700 °C + 1350 °C. The duration of each hold is 6 h. (**e**) Particle size distributions calculated from SEM data. The values presented in the figure (**e**) denote heat-treatment temperatures and time (°C/h).

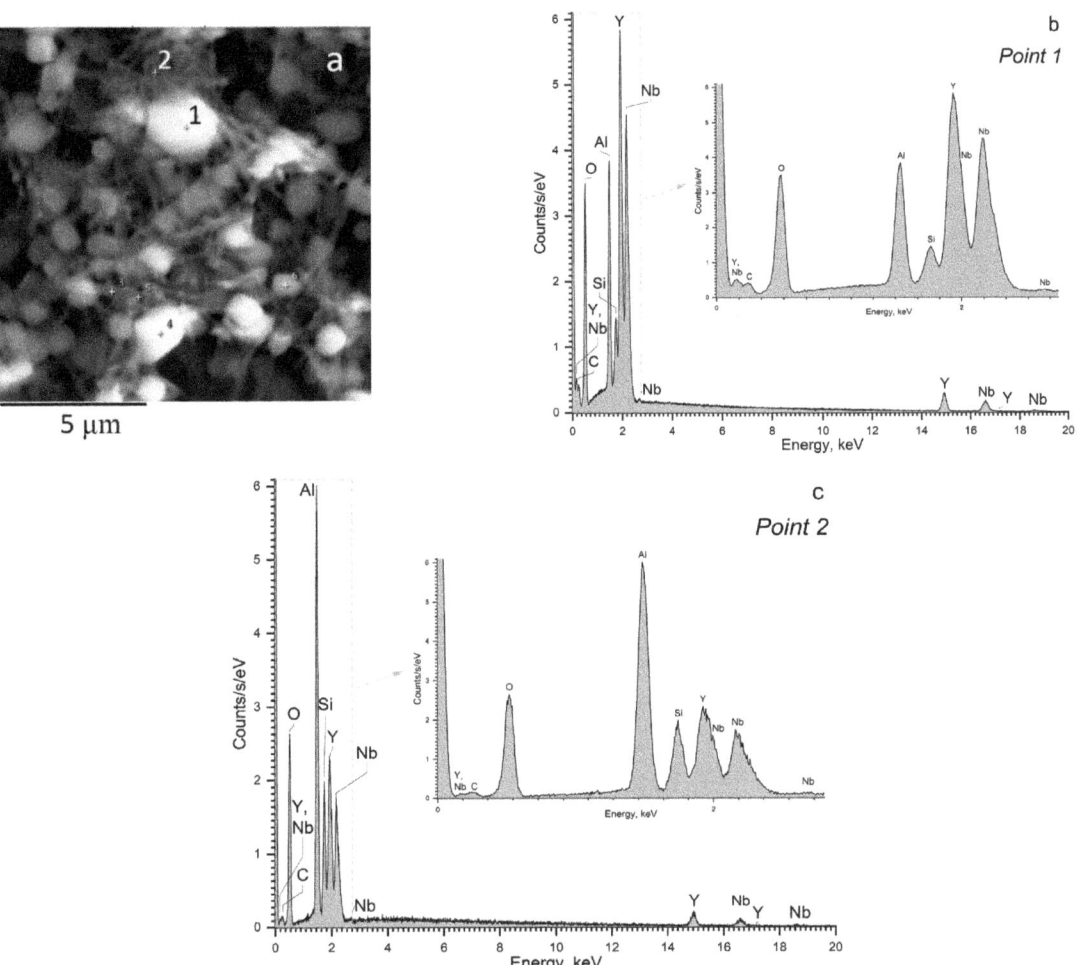

Figure 8. SEM-EDX analysis of the etched surface of the glass-ceramic prepared by the heat-treatment at 700 °C + 1350 °C. The duration of each hold is 6 h: (**a**) SEM image; (**b**) EDX spectrum, point 1; (**c**) EDX spectrum, point 2. The points 1 and 2 are shown in Figure 8a by the corresponding numbers. The numbers and plus sign in Figure 8a show the places from which the EDX spectra were recorded.

According to Figure 8b and Table 2, the bright round-shaped crystals denoted by the number 1 in Figure 8a consist of yttrium, niobium, and oxygen ions. We also conducted EDX analysis of the other round-shaped particles of different sizes and found that they have a very similar elemental composition. Aluminum and silicon ions that appear in the EDX spectrum in Figure 8b probably come from the surrounding matrix and from the rod-like crystals located nearby (we should bear in mind that aluminum oxide is more resistant to hydrofluoric acid than silica). Figure 8c presents the results of the EDX analysis of the material from the point 2 in Figure 8a. Point 2 was chosen as a place where a number of the rod-like crystals are located, see Figure 8a. This composition is enriched in aluminum and silicon and depleted in yttrium and niobium as compared with the composition in the point 1.

Table 2. Compositions of the points 1 and 2 on the etched surface of the glass-ceramic prepared by the heat-treatment at 700 °C + 1350 °C (see Figure 8a).

Spectrum Mark	O	Al	Si	Y	Nb	Total
	% Atom					
1	65.33	11.36	2.72	8.64	11.95	100.00
2	63.31	19.39	6.05	5.99	5.26	100.00

3.5. Absorption Spectra

Absorption spectra of the initial glass and glass-ceramics are determined by absorption of Co^{2+} ions. The spectrum of the initial blue glass shows three wide absorption bands in the visible region at 510, 590, and 650 nm, and a weak broad band in IR region spanning from ~1000 to ~2300 nm with a maximum at ~1460 nm, see Figure 9. The shape of the absorption spectrum remains near unchanged after heat-treatments at 700 °C and 700 + 750 °C for 6 h; these samples are also blue-colored, see Figure 1. When the main crystalline phase of β-quartz ss starts to crystallize at 800 °C, the material becomes violet-colored and its absorption drastically changes. It is characterized by a number of well-resolved intense bands in the visible range of spectrum with maxima at 505, 550, 580, and 595 nm, see Figure 9. There is also a weak inflection at 430 nm. In the IR region, broad bands are observed with maxima around 1250, 1500, 1720, and 1970 nm. The spectrum remains near unchanged up to heat-treatment at 700 + 1000 °C, when the color of the materials changes again to violet-blue, and new spectral features evolve. In the visible region, the spectrum is characterized by a number of narrow bands at 545, 580, and 620 nm. In the IR, there are three wide bands at approximately 1260, ~1400, and 1560 nm.

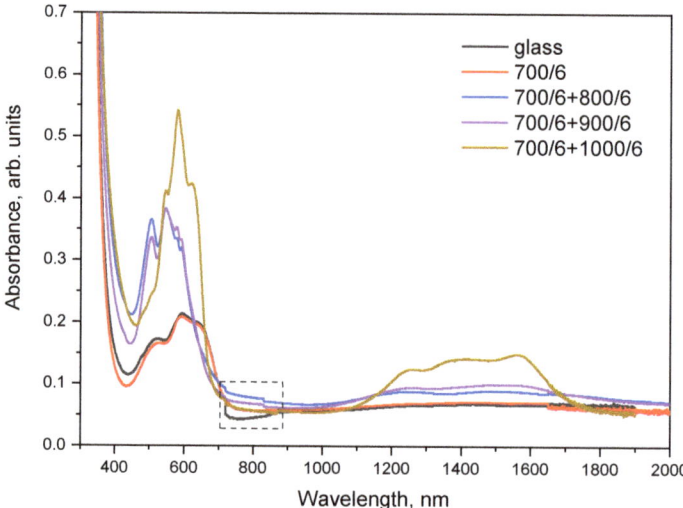

Figure 9. Absorption spectra of the initial and heat-treated glasses. The values presented in the figure denote heat-treatment temperatures and time (°C/h). The samples thickness is ~1 mm. The spectral range marked with the dotted line shows the steps due to the change in detectors.

The spectrum of the initial glass is characteristic for Co^{2+} ions in silicate and aluminosilicate glasses and thoroughly discussed in the literature [23,38–40]. This type of spectrum is caused by the absorption of octahedrally and tetrahedrally coordinated Co^{2+} species with the predominance of the octahedrally coordinated Co^{2+} species (the presence of octahedrally coordinated Co^{2+} ions is confirmed by a characteristic absorption band at 510 nm). This spectrum is similar to the spectrum of the cobalt-doped initial glass of the same composition nucleated by ZrO_2, see Ref. [23]. This similarity suggests that Co^{2+} ions do not enter into the crystalline phase of the nucleating agent but remain in the residual glass phase. The fact that the absorption spectrum of the initial glass does not change with heat-treatment at 700 °C and even at 700 + 750 °C confirms this suggestion because the volume fraction of $YNbO_4$ nanocrystals increases with these heat-treatments, see Figure 3a. The spectral features of absorption spectra of glass-ceramics prepared by the heat-treatments at the second stage at 800, 850, 900, and 1000 °C are similar to those discussed in Ref. [23]. It is not surprising because in Ref. [23], we studied the structural states of Co^{2+} ions upon formation of the same main crystalline phase of β-quartz ss in the glass of the same composition nucleated by ZrO_2. The spectrum of the violet-colored glass-ceramic prepared at 800 and 900 °C, resembles the so-called type II spectrum of Ref. [23]. It is assigned to a combination of the two coordinations, with a predominance of octahedral coordination, which gives rise to a band at 505 nm. Intensities of absorption bands at 505 and 550 nm are relatively high, indicating the possibility of distortion of the octahedral symmetry. The presence of tetrahedral sites is manifested by the bands at 550, 580, and 595 nm. The band at 1220 nm arises from Co^{2+} ions in octahedral sites, while bands at 1260, 1400, and 1570 nm are due to tetrahedral sites of Co^{2+}. Following Bogdanova et al. [40], we state [23] that Co^{2+} ions isomorphously replace Li^+ ions located mostly in octahedral sites in the disordered structure of β-quartz ss crystallized in glass having high viscosity. The appearance of the so-called type III spectrum [23] after the heat-treatment at 1000 °C could be explained by a significant change in the local environment around the Co^{2+} ions in the structure of β-quartz ss. An increase in intensities of the bands at 545, 580, and 620 nm in the visible range of spectrum, and of bands at 1390 and 1570 nm in the IR (Figure 9), manifests an increase in tetrahedrally coordinated Co^{2+} sites. Table 1 of Ref. [23] shows an assignment of the absorption bands of Co^{2+} ions recorded in the initial glass and glass-ceramics.

3.6. Linear Coefficient of Thermal Expansion

The CTE value is ~61.0 ± 1.0 × 10^{-7} K^{-1} for the initial glass and the glass heat-treated at 700 °C and at 700 + 750 °C, which is consistent with their similar phase composition. The CTE value drastically changes upon crystallization of the main phase, β-quartz ss, see Table 3. The lowest CTE value of the transparent glass-ceramics of the LAS system obtained in this study is 8.5 ± 1.0 × 10^{-7} K^{-1}. Crystallization of β-quartz ss in glasses of the LAS system doped with TiO_2, ZrO_2, or their mixture usually results in transparent glass-ceramics with near zero CTE value [1,2]. We believe that the crystals of yttrium orthoniobate with the CTE value of about 100–110 × 10^{-7} K^{-1} [27] are responsible for the obtained CTE values, while the combination of two crystalline phases with different CTE values and of the residual glass phase is responsible for the complex dependence of the CTE value on the heat-treatment schedule. It is worth mentioning that the proper choice of the heat-treatment schedule allows us to obtain glass-ceramics with the CTE varied in the broad range of the values.

Table 3. The CTE values of the initials and the heat-treated glass.

Heat-Treatment Schedule, °C	Glass	700	700 + 750	700 + 800	700 + 850	700 + 900	700 + 1000	700 + 1350
CTE, ±1.0, × 10^{-7} K^{-1}	61.5	60.5	61.5	12.0	8.5	11.5	13.5	27.0

4. Discussion

The initial glass is found to be inhomogeneous. It contains nanosized metastable crystals of T'-$YNbO_4$ with the distorted tetragonal structure. They crystallized from the initial glass during melt casting, cooling, and annealing. The T'-crystals are preserved at room temperature. According to SAXS findings, there is an order in the distribution of these T'-$YNbO_4$ crystals, which confirms the role of yttrium niobate as a nucleating agent. Glass-ceramics obtained by isothermal heat-treatments at 700–900 °C also contain the T'-

phase. In our previous studies [6,7], we have demonstrated that the formation of M-YNbO$_4$ nanocrystals in glass-ceramics typically occurs only after isothermal heat-treatments at temperatures exceeding 950 °C.

According to in situ HT-PXRD study, the T'-phase had broadened peaks. The T-phase with sharp lines developed above 1000 °C. No yttrium niobate other than YNbO$_4$ was observed during the heating process. On cooling, the T-phase was gradually converted to the M-phase by a very sluggish phase transformation below 600 °C. It should be noted the T' to M-phase transformation did not occur; the T'-phase crystallized during heat-treatments was obtained at room temperature.

Thus, the in situ HT-PXRD study revealed that the M-YNbO$_4$ crystals are formed not during isothermal heat-treatments but during cooling the prepared glass-ceramics in the temperature range from 600 °C to room temperature. M-phase is formed by the transformation of crystals of the high-temperature tetragonal phase with a sheelite-like structure, T-YNbO$_4$.

In our studies, a complete T–M phase transition did not occur in the selected heat-treatment schedules, resulting in the coexistence of the M-phase and the T-phase at room temperature. It can be inferred that the stabilization of the tetragonal niobate phase with sheelite structure at room temperature is influenced by the size effect through the contribution of high surface energy to the Gibbs energy of the tetragonal niobate crystals.

Prolonged heat-treatment at 1350 °C ensures that the majority of the niobate phase undergoes crystallization in the form of the T-phase and subsequently transforms into the M-phase during cooling to room temperature.

In this study, we aimed to reveal the mechanism of phase transformations in yttrium orthoniobate nanocrystals upon heating the initial glass and cooling the fabricated glass-ceramic. The glass composition studied in this work was the model one. Relatively high concentrations of yttrium and niobium oxides made it easier for us to study these phase transformations. It should be noted that the crystalline phase of YNbO$_4$ is not only the nucleating agent but also a promising host for other rare-earth ions [6,7]. That is why it is quite natural that the proportion of Y$_2$O$_3$ and Nb$_2$O$_5$ is above the proportions of TiO$_2$/ZrO$_2$ usually used in industry. It should be noted that rare-earth ions do not enter any crystalline phase of LAS glass-ceramics produced in industry.

Doping the glass with small amount of cobalt oxide allowed us to demonstrate that the developed glass-ceramics are multifunctional materials. In this work, we have shown that cobalt ions, as representatives of transition metal ions, do not enter the crystalline phase of yttrium orthoniobate, but selectively enter crystals of the β-quartz ss. According to our previous studies [6,7] rare-earth ions, such as Er^{3+}, Yb^{3+}, and Eu^{3+}, do not enter the structure of the β-quartz ss, but form their own phase of rare-earth orthoniobate or selectively enter the structure of yttrium orthoniobate [6,7]. By this means, rare-earth orthoniobates play the dual role of nucleating agents and active crystalline phases containing rare-earth ions. Thus, we demonstrated the development of transparent glass-ceramics with near zero thermal expansion and with a selective doping of rare-earth and transition metal ions into different crystalline phases. These materials are promising for photonic applications.

5. Conclusions

Transparent glass-ceramics based on nanosized crystals of yttrium orthoniobates, YNbO$_4$, and β-quartz ss and doped with cobalt ions were prepared for the first time. The initial glass and glass-ceramics obtained by isothermal heat-treatments at 700–750 °C contain the only phase of T'-YNbO$_4$ with the distorted tetragonal structure and the crystal size of ~10 nm. With an increase in the heat-treatment temperature, the size and volume fraction of YNbO$_4$ crystals grow; in the process, the crystal structure transforms to the tetragonal one, T-phase. In glass-ceramics obtained by heat-treatment at 1000 °C the transformation to the monoclinic form (M-phase) begins.

The sequence of phase transformations in the yttrium orthoniobates crystals was studied in detail by means of in situ high-temperature X-ray diffraction and differential thermal analysis. It was revealed for the first time that formation of the monoclinic phase of yttrium orthoniobate occurs only from the high-temperature tetragonal phase with a sheelite-like structure upon cooling according to the displacive ferroelastic transformation of a second order.

Nanosized crystals of yttrium orthoniobates act as nucleating agents for bulk crystallization of lithium aluminosilicate solid solutions with β-quartz structure, the main crystalline phase of glass-ceramics, which starts to crystallize at 800 °C.

It was demonstrated that cobalt ions from the initial glass selectively enter the crystals of β-quartz ss. The multifunctional transparent glass-ceramics were developed. In these glass-ceramics, rare-earth ions can selectively enter the nanosized crystals of yttrium niobate, while the transition metal ions can selectively enter the β-quartz ss. The large amount of the main crystalline phase of β-quartz ss ensures the low values of coefficient of thermal expansion and high thermal shock resistance of the developed glass-ceramics. Heat-treatments at different temperatures result in the development of glass-ceramics containing crystalline phases of different structures, thus ensuring a variety of their physical and optical properties.

Author Contributions: Conceptualization, O.D., A.B. and A.Z.; methodology, O.D., A.B., I.A., M.T. and D.D.; software, A.B.; validation, O.D., A.B. and I.A.; investigation, O.D., A.B., I.A., V.G., S.Z. and M.T.; resources, O.D. and D.D.; data curation, S.Z. and K.B.; writing—original draft preparation, O.D. and A.B.; writing—review and editing, O.D.; visualization, A.B.; supervision, O.D.; project administration, O.D.; funding acquisition, O.D. and A.Z. All authors have read and agreed to the published version of the manuscript.

Funding: This research received no external funding.

Institutional Review Board Statement: Not applicable.

Informed Consent Statement: Not applicable.

Data Availability Statement: Data is contained within the article.

Acknowledgments: The SEM study was performed using the analytical equipment of the Engineering Center of the St. Petersburg State Institute of Technology.

Conflicts of Interest: The authors declare no conflict of interest.

References

1. Holand, W.; Beall, G.H. (Eds.) *Glass-Ceramic Technology*, 3rd ed.; Wiley/American Ceramic Society: New York, NY, USA, 2020; pp. 82–87, 273–281.
2. Bach, H.; Krause, D. (Eds.) *Low Thermal Expansion Glass Ceramics*, 2nd ed.; Springer: Berlin/Heidelberg, Germany, 2005; p. 260.
3. Fujita, S.; Tanabe, S. Structural evolution of Er^{3+} ions in Li_2O–Al_2O_3–SiO_2 glass-ceramics. *J. Ceram. Soc. Jpn.* **2008**, *116*, 1121–1125. [CrossRef]
4. Beall, G.H. Milestones in glass-ceramics: A personal perspective. *Int. J. Appl. Glass Sci.* **2014**, *5*, 93–103. [CrossRef]
5. Dymshits, O.; Shepilov, M.; Zhilin, A. Transparent glass-ceramics for optical applications. *MRS Bull.* **2017**, *42*, 200–205. [CrossRef]
6. Dymshits, O.S.; Alekseeva, I.P.; Zhilin, A.A.; Tsenter, M.Y.; Loiko, P.A.; Skoptsov, N.A.; Malyarevich, A.M.; Yumashev, K.V.; Mateos, X.; Baranov, A.V. Structural characteristics and spectral properties of novel transparent lithium aluminosilicate glass-ceramics containing (Er,Yb)NbO_4 nanocrystals. *J. Lumin.* **2015**, *160*, 337–345. [CrossRef]
7. Loiko, P.A.; Dymshits, O.S.; Alekseeva, I.P.; Zhilin, A.A.; Tsenter, M.Y.; Vilejshikova, E.V.; Yumashev, K.V.; Bogdanov, K.V. Structure and spectroscopic properties of transparent glass-ceramics with (Eu^{3+},Yb^{3+}):$YNbO_4$ nanocrystals. *J. Lumin.* **2016**, *179*, 64–73. [CrossRef]
8. Blasse, G. Luminescence processes in niobates with fergusonite structure. *J. Lumin.* **1976**, *14*, 231–233. [CrossRef]
9. Mao, J.; Jiang, B.; Wang, P.; Qiu, L.; Abass, M.T.; Wei, X.; Chen, Y.; Yin, M. A study on temperature sensing performance based on the luminescence of Eu^{3+} and Er^{3+} co-doped $YNbO_4$. *Dalton Trans.* **2020**, *49*, 8194–8200. [CrossRef]
10. Wang, X.; Li, X.; Shen, R.; Xu, S.; Zhang, X.; Cheng, L.; Sun, J.; Zhang, J.; Chen, B. Optical transition and luminescence properties of Sm^{3+}-doped $YNbO_4$ powder phosphors. *J. Am. Ceram.* **2020**, *103*, 1037–1045. [CrossRef]
11. Hirano, M.; Ishikawa, K. Intense up-conversion luminescence of Er^{3+}/Yb^{3+} co-doped $YNbO_4$ through hydrothermal route. *J. Photochem. Photobiol. A Chem.* **2016**, *316*, 88–94. [CrossRef]
12. Đačanin, L.R.; Lukić-Petrović, S.R.; Petrović, D.M.; Nikolić, M.G.; Dramićanin, M.D. Temperature quenching of luminescence emission in Eu^{3+}- and Sm^{3+}-doped $YNbO_4$ powders. *J. Lumin.* **2014**, *151*, 82–87. [CrossRef]

13. Niu, C.; Li, L.; Li, X.; Lv, Y.; Lang, X. Upconversion photoluminescence properties of Ho^{3+}/Yb^{3+} co-doped YNbO$_4$ powder. *Opt. Mater.* **2018**, *75*, 68–73. [CrossRef]
14. Đačanin Far, L.; Lukić-Petrović, S.R.; Đorđević, V.; Vuković, K.; Glais, E.; Viana, B.; Dramićanin, M.D. Luminescence temperature sensing in visible and NIR spectral range using Dy^{3+} and Nd^{3+} doped YNbO$_4$. *Sens. Actuator A Phys.* **2018**, *270*, 89–96. [CrossRef]
15. Đačanin Far, L.; Ćirić, A.; Sekulić, M.; Periša, J.; Ristić, Z.; Antić, Ž.; Dramićanin, M.D. Judd-Ofelt description of radiative properties of YNbO$_4$ activated with different Eu^{3+} concentrations. *Optik* **2023**, *272*, 170398.
16. Zhou, Y.; Ma, Q.; Lü, M.; Qiu, Z.; Zhang, A. Combustion synthesis and photoluminescence properties of YNbO4-based nanophosphors. *J. Phys. Chem. C* **2008**, *112*, 19901–19907. [CrossRef]
17. Barth, T. The structure of synthetic, metamict, and recrystallized fergusonite. *Nor. Geol. Tidsskr.* **1926**, *9*, 23–36.
18. Komkov, A.I. The structure of natural fergusonite, and of a polymorphic modification. *Kristallografiya (Sov. Phys. Crystallogr.)* **1959**, *4*, 836–841. (In Russian)
19. Stubican, V. High-temperature transitions in rare-earth niobates and tantalates. *J. Am. Ceram. Soc.* **1964**, *47*, 55–58. [CrossRef]
20. Jurkschat, K.; Sarin, P.; Siah, L.F.; Kriven, W.M. In Situ High-temperature phase transformations in rare earth niobates. *Adv. X-ray Anal.* **2004**, *47*, 357–359.
21. Yamaguchi, O.; Matsui, K.; Kawabe, T.; Shimizu, K. Crystallization and transformation of distorted tetragonal YNbO$_4$. *J. Am. Ceram. Soc.* **1985**, *68*, 275–276. [CrossRef]
22. Mather, S.A.; Davies, P.K. Nonequilibrium phase formation in oxides prepared at low temperature: Fergusonite-related phases. *J. Am. Ceram. Soc.* **1995**, *78*, 2737–2745. [CrossRef]
23. Kang, U.; Dymshits, O.S.; Zhilin, A.A.; Chuvaeva, T.I.; Petrovsky, G.T. Structural states of Co(II) in β-eucryptite-based glass-ceramics nucleated with ZrO$_2$. *J. Non-Cryst. Solids* **1996**, *204*, 151–157. [CrossRef]
24. Lipson, H.; Steeple, H. *Interpretation of X-ray Powder Patterns*; McMillan, Ed.; Martins Press: London, UK, 1970; p. 335.
25. Filipovich, V.N. Toward the theory of Small-Angle X-ray Scattering. *Zh. Tekh. Fiz.* **1956**, *26*, 398–416. (In Russian)
26. Andreev, N.S.; Filipovich, V.N.; Mazurin, O.V.; Porai-Koshits, E.A.; Roskova, G.P. *Phase Separation in Glass*; Elsevier Science Publishers B.V.: Amsterdam, The Netherlands, 1984.
27. Wu, F.; Wu, P.; Zhou, Y.; Chong, X.; Feng, J. The thermo-mechanical properties and ferroelastic phase transition of RENbO$_4$ (RE = Y, La, Nd, Sm, Gd, Dy, Yb) ceramics. *J. Am. Ceram. Soc.* **2020**, *103*, 2727–2740. [CrossRef]
28. Bondar, A.; Koroleva, L.N.; Toropov, N.A. Phase equilibriums in the yttrium sesquioxide-niobium pentoxide system. *Izv. Akad. Nauk SSSR Neorg. Mater.* **1969**, *5*, 1730–1733.
29. Moore, R.L.; Haynes, B.S.; Montoya, A. Effect of the local atomic ordering on the stability of β-spodumene. *Inorg. Chem.* **2016**, *55*, 6426–6434. [CrossRef] [PubMed]
30. Trunov, V.K.; Efremov, V.A.; Velikopodnyi, Y.A.; Averina, I.M. The structure of YNbO$_4$ crystals at room temperature. *Kristallografiya (Sov. Phys. Crystallogr.)* **1981**, *26*, 67–71. (In Russian)
31. McMillan, P. Structural studies of silicate glasses and melts—Applications and limitations of Raman spectroscopy. *Am. Min.* **1984**, *69*, 622–644.
32. Huanxin, G.; Zhongcai, W.; Shizhuo, W. Properties and structure of niobosilicate glasses. *J. Non-Cryst. Solids* **1989**, *112*, 332–335. [CrossRef]
33. Yashima, M.; Lee, J.-H.; Kakihana, M.; Yoshimura, M. Raman spectral characterization of existing phases in the Y$_2$O$_3$-Nb$_2$O$_5$ system. *J. Phys. Chem. Solids* **1997**, *58*, 1593–1597. [CrossRef]
34. Alekseeva, I.; Dymshits, O.; Ermakov, V.; Zhilin, A.; Petrov, V.; Tsenter, M. Raman spectroscopy quantifying the composition of stuffed β-quartz derivative phases in lithium aluminosilicate glass-ceramics. *J. Non-Cryst. Solids* **2008**, *354*, 4932–4939. [CrossRef]
35. Blasse, G. Vibrational spectra of yttrium niobate and tantalate. *J. Solid State Chem.* **1972**, *7*, 169–171. [CrossRef]
36. Guene-Girard, S.; Courtois, J.; Dussauze, M.; Heintz, J.-M.; Fargues, A.; Roger, J.; Nalin, M.; Cardinal, T.; Jubera, V. Comparison of structural and spectroscopic properties of Ho^{3+}-doped niobate compounds. *Mater. Res. Bull.* **2021**, *143*, 111451. [CrossRef]
37. Sekulić, M.; Dramićanin, T.; Ćirić, A.; Đačanin Far, L.; Dramićanin, M.D.; Đorđević, V. Photoluminescence of the Eu^{3+}-activated Y$_x$Lu$_{1-x}$NbO$_4$ (x = 0, 0.25, 0.5, 0.75, 1) solid-solution phosphors. *Crystals* **2022**, *12*, 427. [CrossRef]
38. Weyl, W.A. *Coloured Glasses*, 2nd ed.; Dawson's of Pall Mall: London, UK, 1959; p. 558.
39. Bamford, C.R. The application of ligand field theory to coloured glasses. *Phys. Chem. Glasses* **1962**, *3*, 189–202.
40. Bogdanova, G.C.; Antonova, C.L.; Dzhurinskii, B.F. Distribution of coloring ions in the structure of glass-ceramics. *Izv. Akad. Nauk SSSR Neorg. Mater.* **1969**, *5*, 204–206, (in *J. Am. Ceram. Soc.* **1971**, *54*, 193d).

Disclaimer/Publisher's Note: The statements, opinions and data contained in all publications are solely those of the individual author(s) and contributor(s) and not of MDPI and/or the editor(s). MDPI and/or the editor(s) disclaim responsibility for any injury to people or property resulting from any ideas, methods, instructions or products referred to in the content.

Article

Tuning the Coefficient of Thermal Expansion of Transparent Lithium Aluminosilicate Glass-Ceramics by a Two-Stage Heat Treatment

Andrey S. Naumov *, Georgiy Yu. Shakhgildyan, Nikita V. Golubev, Alexey S. Lipatiev, Sergey S. Fedotov, Roman O. Alekseev, Elena S. Ingat'eva, Vitaliy I. Savinkov and Vladimir N. Sigaev

Department of Glass and Glass-Ceramics, Mendeleev University of Chemical Technology, 125047 Moscow, Russia
* Correspondence: naumov.a.s@muctr.ru

Abstract: Transparent glass-ceramics with a Li_2O–Al_2O_3–SiO_2 (LAS) system have been extensively utilized in optical systems in which thermal stability is of utmost importance. This study is aimed to develop thermal treatment routes that can effectively control the structure of transparent LAS glass-ceramics and tune its thermal expansion coefficient within a wide range for novel applications in photonics and integrated optics. The optimal conditions for the nucleation and crystallization of LAS glass were determined by means of differential scanning calorimetry and a polythermal analysis. XRD, Raman spectroscopy, and TEM microscopy were employed to examine the structural changes which occurred after heat treatments. It was found that the second stage of heat treatment promotes the formation of β-eucryptite-like solid solution nanocrystals, which enables effective control of the coefficient of thermal expansion of glass-ceramics in a wide temperature range of −120 to 500 °C. This work provides novel insights into structural rearrangement scenarios occurring in LAS glass, which are crucial for accurately predicting its crystallization behavior and ultimately achieving transparent glass-ceramics with desirable properties.

Keywords: transparent glass-ceramics; zero CTE; low expansion materials; nucleation; crystallization

Citation: Naumov, A.S.; Shakhgildyan, G.Y.; Golubev, N.V.; Lipatiev, A.S.; Fedotov, S.S.; Alekseev, R.O.; Ingat'eva, E.S.; Savinkov, V.I.; Sigaev, V.N. Tuning the Coefficient of Thermal Expansion of Transparent Lithium Aluminosilicate Glass-Ceramics by a Two-Stage Heat Treatment. *Ceramics* **2024**, *7*, 1–14. https://doi.org/10.3390/ceramics7010001

Academic Editors: Narottam P. Bansal and Gilbert Fantozzi

Received: 1 October 2023
Revised: 18 December 2023
Accepted: 20 December 2023
Published: 22 December 2023

Copyright: © 2023 by the authors. Licensee MDPI, Basel, Switzerland. This article is an open access article distributed under the terms and conditions of the Creative Commons Attribution (CC BY) license (https://creativecommons.org/licenses/by/4.0/).

1. Introduction

Transparent glass-ceramics in the Li_2O–Al_2O_3–SiO_2 (LAS) system have been demonstrated to have notable importance in the field of optical instrument engineering, particularly in applications in which thermal stability plays a crucial role [1,2]. This is primarily attributed to their extremely low coefficient of thermal expansion (CTE) and their manufacturability, which allows for the production of large material blanks with outstanding optical homogeneity. Such materials have provided a significant impetus to the development of such areas as astronomy [3,4], ultra-precision metrology [5], the fabrication of navigation devices [6,7], and novel nanometer precision manufacturing techniques, such as extreme ultraviolet lithography [8,9]. The unique thermo-mechanical characteristics exhibited by LAS glass-ceramics can be ascribed to the carefully regulated process of crystallization, facilitating the formation of β-eucryptite $LiAlSiO_4$ solid solutions [10,11]. The β-eucryptite crystal phase exhibits a negative average CTE along the *c*-axis, with values reaching as low as −6.2 ppm/K in the temperature range of 20–1000 °C [12,13]. By combining the thermal expansion of the crystalline phase with the residual amorphous glass phase, LAS glass-ceramics can be engineered to achieve near-zero CTE values, resulting in thermally stable materials. On the other hand, the ability to finely adjust thermomechanical properties through the temperature treatment of initial glasses shows potential for novel applications of LAS glass-ceramics in photonics and integrated optics [14–16]. However, this also calls for more comprehensive studies of the phase separation processes within the LAS glass-forming system.

The controlled crystallization method for LAS glass-ceramics is not unique and typically involves two consecutive stages of glass thermal treatment: (i) nucleation, during which crystalline nuclei form within the amorphous phase; and (ii) crystallization, when the desired crystalline phase grows at the sites of nucleus formation [17–19]. Despite the methodological simplicity of obtaining LAS glass-ceramics, there are numerous factors that influence the properties of the final material, including the chemical composition of the initial glass, the presence of nucleation agents and their quantity, the synthesis method of the glass, and subsequent thermal treatments [20,21]. In this regard, LAS glass-ceramics are involved in a wide scope of research, including investigations into the effect of different types and amounts of nucleating agents [22–24], the determination of nucleation mechanisms and crystallization kinetics [25–27], and the development of various heat-treatment routes to initiate crystal growth [28].

Burgeoning interest in the spatially selective modification of materials through the use of femtosecond laser pulses has illuminated another potential application for transparent glass-ceramics [29]. Direct ultrafast laser writing has opened up new possibilities for the creation of 3D integrated optical and photonic components and devices [30]. Previous research in this field has demonstrated the success of the spatially selective crystallization of nonlinear optical crystals, metallic nanoparticles, and quantum dots [31–33]. The object under investigation in this area of research is glass-ceramics with unique thermal and mechanical properties. Recently, the direct laser writing of functional structures, such as waveguides and directional couplers, in glass-ceramics was demonstrated [15,16]. It is believed that the use of LAS glass-ceramics would enable the production of integrated optical devices and the miniaturization of large optical systems used in space astronomy and navigation, for which size and weight parameters are crucial [14,16,34]. For the fabrication of temperature-insensitive waveguides, not only the CTE of the material but also the temperature dispersion of its refractive index should be taken into account. Thus, the tuning of glass-ceramics' CTE is required, which can help to partially compensate for the change in the refractive index within the temperature. In line with that, the optical transparency of glass-ceramics must remain high, because light scattering can lead to increased optical losses [35].

For the first time, this study aims to explore the potential of wide-range tuning the structure and properties of LAS glass-ceramics, for which we previously demonstrated the possibility of writing integrated optical waveguides using femtosecond laser radiation [16]. The outcomes of this research could pave the way for the development of transparent LAS glass-ceramics with enhanced properties, featuring specified CTE values, and enabling the subsequent creation of complex photonic structures within these materials. This could simplify the manufacturing process of integrated optical devices and facilitate the miniaturization of large optical systems.

2. Materials and Methods
2.1. Glass Synthesis and Crystallization

Reagent grade raw materials were weighed, mixed, and used for the preparation of the batch, which was calculated to produce 500 g of glass. Reagent grade Li_2CO_3 (Lanhit, Moscow, Russia), $Al(OH)_3$ (Labhimos-S, Moscow, Russia), SiO_2 (Lanhit, Moscow, Russia), $Al(PO_3)_3$ (Ekotek, Moscow, Russia), $MgCO_3$ (Spectr-Him, Moscow, Russia), $Ba(NO_3)_2$ (Reahim, Moscow, Russia), TiO_2 (Lanhit, Moscow, Russia), ZrO_2 (Lanhit, Moscow, Russia), ZnO (Spectr-Him, Moscow, Russia), Sb_2O_3 (CT Lantan, Moscow, Russia), and As_2O_3 (Himpromkomplekt, Penza, Russia) were used as raw materials. Batch composition of LAS glass is given in Table 1. The same glass composition was previously used in a work studying the direct laser writing of depressed-cladding optical waveguides in glass-ceramics [16]. The glass batch was loaded into the corundum crucible and placed in the chamber of the bottom-loading electrical furnace equipped with $MoSi_2$ heating elements (Promtermo Ltd., Moscow, Russia). The furnace was heated up to 1600 °C and kept at this temperature for 4 h to homogenize and refine the glass melt. After that, the melt was poured into the preheated mold, which was subsequently transferred into the electric muffle furnace (Termokeramika Ltd., Moscow, Russia) for annealing of the glass cast at 600 °C for 4 h. The fabricated glass block of the size ~150 mm × 90 mm × 15 mm shown in Figure S1 was visually transparent and homogeneous.

Table 1. Batched and analyzed compositions as well as some properties of synthesized glass.

	Glass Composition											
Oxides	SiO_2	Al_2O_3	Li_2O	P_2O_5	TiO_2	ZrO_2	ZnO	MgO	BaO	CaO	As_2O_3	Sb_2O_3
Batch composition, mol. %	61.1	15.9	11.1	4.9	2.1	0.9	0.4	2.0	0.8	0.4	0.2	0.2
Analytical composition, mol. %	60.0	19.2	10.1	4.6	1.9	0.8	0.4	1.9	0.7	0.3	0.1	0.1
	Glass properties											
Glass transition temperature T_g, °C	Crystallization peak T_P, °C		$T_P - T_g$, °C		$CTE_{20\ldots500\,°C}$, ppm/K		Density, g/cm³			Refractive index (n_D)		
647	867		220		5.17		2.46			1.52		

During the glass melting process, the composition of the glass can undergo changes as a result of the evaporation of volatile components and partial dissolution of the crucible. To evaluate the chemical composition of the glass after the synthesis, a 5800 VDV ICP-OES inductively coupled plasma optical emission spectrometer equipped with an Ar flow atmosphere plasma sample injection system (Agilent Technologies Inc., Santa Clara, CA, USA) was employed. The instrumental settings for this technique and the measured elements are presented in Table S1. Complete dissolution of the powdered glass sample was achieved through a combination of HF/HCl/HNO₃ acids. The results of the chemical analysis are presented in Table 1 and indicate good agreement between calculated and analyzed glass composition. Mass losses during the glass melting did not exceed 1.5%. The slight increase in Al_2O_3 content is presumably due to degradation of the corundum crucible.

The glass block was cut into ~10 mm × 10 mm × 2 mm plates and subsequently heat-treated in a muffle furnace with different regimes to produce the glass-ceramics. Polythermal crystallization of the bulk glass samples was also performed using the technique described previously [35].

2.2. Glass and Glass-Ceramics Characterization

The densities of glass and glass-ceramics were measured at room temperature by the Archimedes method with distilled water as an immersing liquid. Bulk samples of initial and nucleated glasses in the form of thin disks (15 ± 1 mg) were used for thermal analysis carried out using the Netzsch STA 449 F3 Jupiter simultaneous thermal analyzer (NETZSCH, Waldkraiburg, Germany). The heating of samples was performed in a Pt crucible at 10 °C/min rate in Ar flow atmosphere. The glass transition temperature (T_g) was determined as the extrapolated onset of the transition, while the crystallization peak temperature (T_P) was defined as the peak extremum temperature in differential scanning calorimetry (DSC) curves. Length changes (ΔL) during heating of the bulk glass and glass-ceramic samples with a size of 20 mm × 4 mm × 4 mm were investigated using a Netzsch DIL 402 SE dilatometer (NETZSCH, Waldkraiburg, Germany), which was equipped with a cooling system allowing for measurements in two temperature ranges: from −180 °C to 25 °C and from 25 °C to 500 °C at a heating rate of 5 °C/min. The CTE values calculated from the elongation curves of $\Delta L / L_0 = f(T)$ function are valid up to 10^{-8} K^{-1}.

Powder X-ray diffraction (XRD) analysis of the initial glass and glass-ceramics was carried out on a D2 Phaser diffractometer (Bruker, Germany) employing nickel-filtered CuKα radiation in 2θ range from 10° to 60°. Crystalline phases were identified by comparing the peak positions and relative intensities in XRD patterns with the ICDD PDF-2 database (released in 2011). The crystallite size was estimated from broadening of the XRD peak at about 26°, without correction for the instrumental broadening, according to Scherrer's equation as follows:

$$D = \frac{K\lambda}{\Delta \cos\theta},\qquad(1)$$

where λ is the wavelength of the X-ray radiation (1.5406 Å), θ is the diffraction angle, Δ is the width of the peak at half of its maximum and K is the constant assumed to be 1 [36]. The crystallized fraction was evaluated as $100 \cdot (A_p/A_x)$, where A_x and A_p are, respectively, the area of the whole XRD pattern (without background) and the area of the peaks considered as the area outside of the broad amorphous XRD pattern (Figure S2). The instrumental background profile was determined from the X-ray scan of an empty silicon-made low-background specimen holder. Indicated areas were calculated (in cps × degrees) using DIFFRAC.EVA software version 5.0.

NTegra Spectra Raman spectrometer (NT-MDT Co., Zelenograd, Moscow, Russia) with an Ar laser beam (with a 488 nm excitation wavelength) was used to record Raman spectra from the studied samples in the form of polished plates. Bulk samples of the obtained glass-ceramics were subjected to investigation by transmission electron microscopy (TEM) using Tecnai Osiris transmission electron microscope (Thermo Fisher Scientific, Waltham, MA, USA) equipped with a spherical aberration corrector. TEM images were obtained in 200 kV mode. To calculate the particle size distribution, 150 particles from the TEM image were analyzed using the ImageJ software version 1.53t.

The refractive index (n_D) of the initial glass was measured with an Abbe DR-M4 refractometer (ATAGO Co., Ltd., Tokyo, Japan) at 589 nm at ~25 °C. The transmission optical spectra of the fabricated samples with thickness of 2 mm were recorded in the visible spectral range using UV-3600 spectrophotometer (Shimadzu, Kyoto, Japan).

3. Results and Discussion

The DSC curve of the initial glass (Figure 1a) shows a single exothermic peak, which is attributed to the crystallization of the β-quartz solid solution, typical for LAS glass compositions in the investigated temperature range [37]. Only the broad halo centered at around 23° was observed in the XRD pattern of the initial glass (Figure 1b), pointing to the absence of any crystalline phase. The main properties of the LAS glass are presented in Table 1.

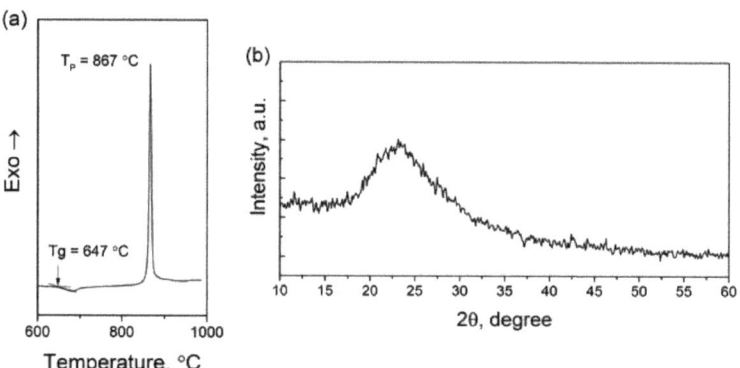

Figure 1. (a) DSC curve and (b) XRD pattern of the synthesized glass.

The attainment of zero CTE values in LAS glass-ceramics is the outcome of an extensive and systematic investigation into the crystallization characteristics of LAS glasses [38,39]. The crystalline phase content that allows us to maintain the transparency of glass-ceramics is in the range from approximately 50% to 90%, in which the crystals are a few tens of nanometers [40]. To achieve a high level of crystallinity, nucleating agents are introduced into the initial glass composition. Additionally, an appropriate temperature treatment route must be adjusted to optimize the kinetics of the glass crystallization process, ensuring the desired ratio between the nucleation and crystal growth rates [9,40].

It is known that a two-stage crystallization approach is generally employed to obtain a fine nanocrystalline structure in transparent glass-ceramics. The first stage promotes the formation of a maximum number of crystal nuclei, while the second stage facilitates the growth of the primary crystalline phase. The nucleation stage has the most significant influence on the ultimate structure and uniform distribution of the crystalline phase throughout the glass-ceramic material [41]. So, indirect methods of monitoring the changes in the physical properties of the material strongly connected with nucleation and crystal growth were employed in this study to ascertain the nucleation temperature of the LAS glass.

3.1. Determination of the Nucleation Temperature

The technique chosen for the evaluation of the nucleation temperature during the production of glass-ceramics was proposed by A. Marotta et al. [42]. This technique involves determining the heat treatment conditions that result in the downward shift of the exothermic peak on the DSC curve, which indicates the facilitation of the crystallization process through the formation of nuclei during the glass pretreatment stage. The most effective nucleation temperature, according to this technique, is determined by monitoring the peak on the curve of the $(T'_P - T_P)$ parameter versus the glass heat treatment temperature. The $(T'_P - T_P)$ parameter represents the difference between the maximum exothermic temperature values for the initial glass (T'_P) and the glass subjected to nucleation pretreatment (T_P).

The DSC curves of the glasses pre-treated near the Tg for a fixed holding time of 2 h are presented in Figure 2a. The evident extremum of $(T'_P - T_P)$ is observed due to the non-linear shift of the position of the exothermic peak under the increase in the heat treatment temperature. The temperature corresponding to this extremum was selected as the nucleation temperature for the fabricated LAS glass.

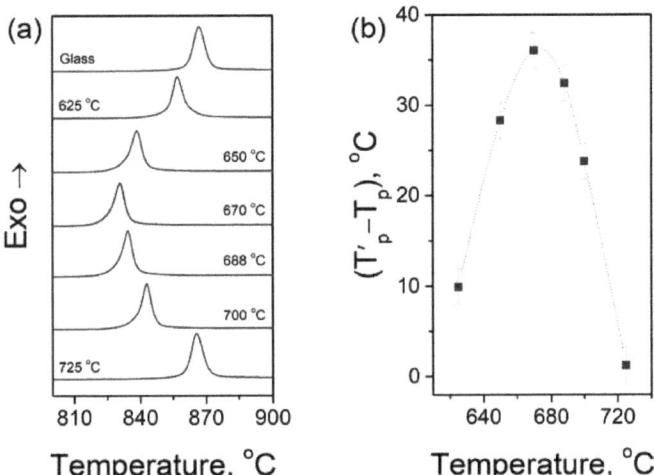

Figure 2. (a) DSC curves (vertically shifted for clarity) of the initial glass and after nucleation pretreatment at the indicated temperatures for 2 h; (b) Shift of the crystallization peak temperature T_P in pretreated samples (with respect to T_P' of the initial glass) vs. nucleation pretreatment temperature.

To optimize the holding time at the nucleation temperature, a comparable method was used, which includes assessing the maximum displacement of the $(T'_P - T_P)$ parameter in relationship to the holding time [42]. The DSC curves of the glasses after treatments at 670 °C for various holding times are presented in Figure 3a. Our analysis of the $(T'_P - T_P)$ dependence reveals that a plateau is reached after a holding time of 2 h. This suggests that the formation of new crystallization nuclei reaches its maximum and does not continue to increase with longer holding times. A similar phenomenon has been observed in studies on the formation and evolution of nanoinhomogeneities in lithium and magnesium aluminosilicate glasses using small-angle neutron scattering [18,43]. Consequently, a holding time of 2 h is believed to be sufficient for the nucleation of synthesized glass, ensuring the formation of the largest possible number of crystal nuclei. The data on the crystallization peak temperatures are summarized in Table S2.

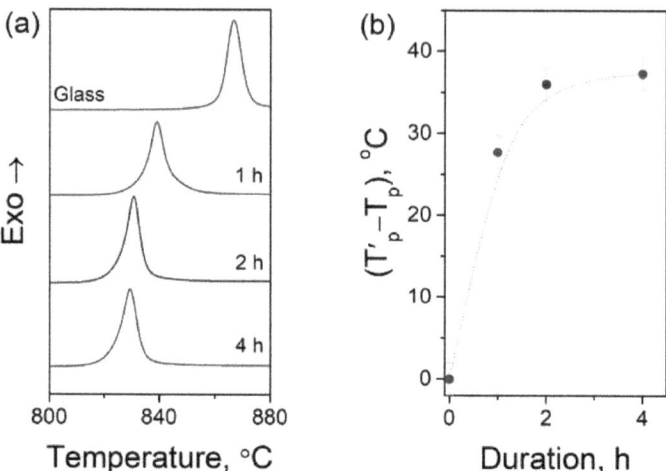

Figure 3. (a) DSC curves (vertically shifted for clarity) of initial glass and after nucleation pretreatment at 670 °C for the indicated duration; (b) Shift of the crystallization peak temperature T_P in pretreated samples (with respect to T_P' of the initial glass) vs. nucleation pretreatment duration.

Hence, the nucleation process for obtaining transparent glass-ceramics in our study involves the pre-treatment of the glass at 670 °C for 2 h. A comparison of the XRD patterns and Raman spectra of the initial glass with those of the glasses after the nucleation pre-treatment indicates no visible changes (Figure S3). However, previous studies on LAS glasses with similar compositions revealed that during nucleation, $ZrTiO_4$ nanocrystals with sizes around a few nm form in glass [44,45]. These nanocrystals serve as a platform for the epitaxial growth of crystal phases based on quartz and lithium aluminosilicates [46,47]. The small size and limited number of crystal nuclei make it challenging to detect them using common techniques such XRD and Raman spectroscopy. Nevertheless, modern TEM methods can provide detailed insights into the mechanism of the formation of such crystal nuclei. It was established that the nucleation mechanism involves liquid–liquid phase separation, resulting in the formation of TiO_2- and ZrO_2-enriched zones, followed by the formation of $ZrTiO_4$ nanocrystals with a core-shell structure [46–48]. The shell consists of an amorphous phase enriched in Al_2O_3, which further inhibits the growth of $ZrTiO_4$. The optimized temperature–time regime of nucleation allows us to move forward in studying the possibilities of the variation in the structure and properties of glass-ceramics by the control of crystallization conditions during the second stage of heat treatment.

3.2. Effect of Crystallization Conditions on the Microstructure and Properties of Glass-Ceramics

A polythermal analysis in the gradient furnace was performed for glass samples to obtain a preliminary view of the crystallization behavior. The main aim of this analysis was to identify the temperature range at which the glass-ceramics maintain their transparency after crystal growth [35]. The hot end temperature of the gradient furnace was varied between 800 and 1000 °C, while the holding times used were 1 or 6 h. It was observed that at temperatures above 900 °C, intense crystallization took place, resulting in the complete opacity of the glass-ceramics. However, when the hot end temperature was reduced to

800 °C, three distinct zones (transparent, opalescent, and completely opaque) can be seen in the samples. No visible changes were observed up to 710 °C, but higher temperatures led first to the appearance of strong opalescence and then to a loss of transparency in the material. It is important to note that the described polythermal analysis was performed on the initial glass samples without a nucleation pre-treatment, resulting in the uncontrolled growth of the crystalline phase. This caused a loss of transparency and the cracking of the glass-ceramic samples at temperatures above 720 °C, which is more than 100 °C lower than the temperature of the exopeak maximum for glass (T_P = 867 °C). Thus, it is impractical to use only the temperature of the crystallization peak for the production of transparent glass-ceramics since the excessively high growth rate of crystals makes it difficult to control the nanocrystals' size [26].

A series of heat treatments in the temperature range of 690–730 °C for a duration of 20 h were carried out on the pre-nucleated glass samples. The results obtained from the polythermal analysis indicated the presence of both transparent and opalescent glass-ceramics in this temperature range (Figure 4). A visual examination of the samples revealed their cracking at temperatures exceeding 720 °C. The cracking occurs due to microstresses arising from the difference in CTEs between the precipitated crystalline phase and residual glass phase (−6.2 ppm/K and 6.0 ppm/K, respectively) when their magnitude exceeds the material strength [49,50]. Therefore, a temperature of 710 °C was selected as the optimal temperature for crystal growth during the second stage of crystallization, resulting in the production of transparent glass-ceramics with no visible defects.

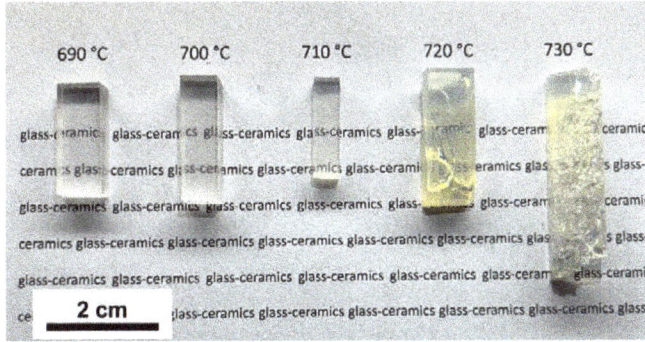

Figure 4. Photo of glass samples obtained after the nucleation at 670 °C for 2 h and crystallization in the range of 690–730 °C for 20 h.

The duration of the crystal growth stage has a direct impact on the final structure and therefore the properties of the synthesized glass-ceramics. Thus, the next series of two-stage heat treatments included a holding time of the second stage up to 30 h. Further, the samples of the glass-ceramics are designated as GC-X, where X is the duration of the second stage of crystallization at 710 °C. Figure 5a presents the Raman spectra of the initial glass and glass-ceramic samples after the performed heat treatments. The spectra exhibit two distinct bands: band A in the range of 470–480 cm^{-1} and band B in the range of 1070–1120 cm^{-1}. Variations in the holding time of the second stage of the heat treatment result in shifts in the position of the maxima of both bands. Notably, band A demonstrates a significant increase in intensity, whereas the increase in band B is less pronounced. Band A registered for the initial glass reaches its maximum at 472 cm^{-1}, indicating transverse vibration modes [51] associated with the Si–O–Si bond of the bridging oxygen during the bonding of [SiO$_4$] tetrahedrons. Additionally, band B in the original glass is observed at 1073 cm^{-1}, representing the antisymmetric stretching vibration of the Si–O bond within the Q^n structures, where n denotes the number of bridging oxygens [52]. As the exposure time exceeds 22.5 h, band A becomes narrower and its maximum shifts to the region of

480 cm^{-1}. This shift may be attributed to a decrease in the Si–O–Si bond angles caused by the crystallization of the silicate phase.

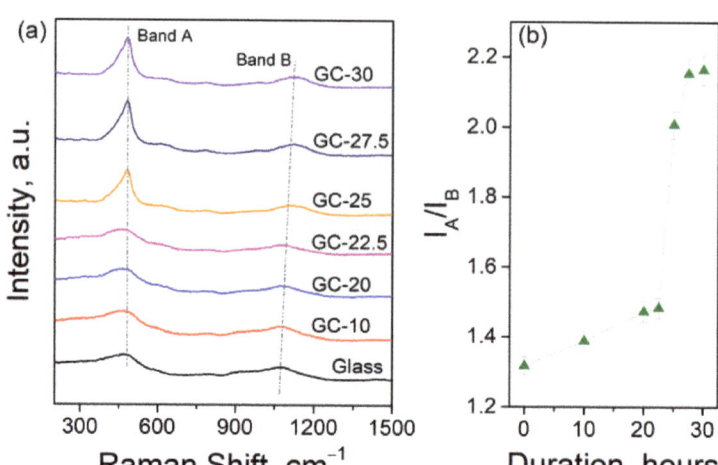

Figure 5. (**a**) Raman spectra of the glass and glass-ceramic samples; (**b**) Intensity ratio of band A and B vs. the crystallization duration.

The shift of the maximum of band B to the region of 1120 cm^{-1}, which is also observed with the increasing holding time of the heat treatment, may be attributed to the enhanced binding of the residual silicate glass phase. This can be explained by the fact that a significant portion of the Li$_2$O glass modifier enters the crystalline phase, resulting in a reduction in non-bridging oxygen in the glass network. Previous studies have associated the bands in the mentioned regions in the Raman spectra of crystalline phases in the LAS system with SiO$_2$-β-quartz, LiAlSi$_2$O$_6$, and LiAlSiO$_4$-β-eucryptite phases. However, due to the overlapping of the bands and their broadening in glasses, it is challenging to provide an accurate interpretation of the phases. Nevertheless, the obtained data tentatively indicate that the observed change in the Raman spectra is related to the crystallization of β-quartz solid solutions, which is typical for LAS glasses.

To describe the scenario of the temporal evolution of crystalline phases in the samples during the second stage of the heat treatment, we examined the variations in the I$_A$/I$_B$ intensity ratios, as depicted in Figure 5b. It is evident that at the duration of up to 22.5 h, the I$_A$/I$_B$ ratio gradually increases. However, at 25 h, the stepped, nearly twofold increase in I$_A$/I$_B$ was found with further plateaus at 27.5 and 30 h. This distinctive pattern indirectly signifies a possible significant change in the thermomechanical characteristics of the glass-ceramics.

The XRD patterns of the glass-ceramic samples, obtained after the nucleation and subsequent crystallization at a temperature of 710 °C with varying holding times, are presented in Figure 6a. It is noteworthy that all the XRD patterns exhibit distinct reflexes, whose position remains virtually unchanged as the holding time of the heat treatment increases. The longer the holding time, the higher the intensity of the reflexes. The primary XRD peak, located at an angle of 2θ = 25.7°, could potentially correspond to multiple crystalline phases, as displayed in Figure 6b. It is plausible that the predominant crystalline phase formed in this glass under the applied heat treatment conditions is a β-quartz solid Li$_2$O·Al$_2$O$_3$·nSiO$_2$-type solution. The diffraction peaks of this solid solution are shifted by approximately 0.5–1° towards larger angles compared to the reflections of β-eucryptite (PDF card #00-026-0839), with the magnitude of deviation being proportional to the angle. This shift can be attributed to the non-stoichiometric SiO$_2$ content in the solid solution and due to the complex multicomponent composition of the initial glass.

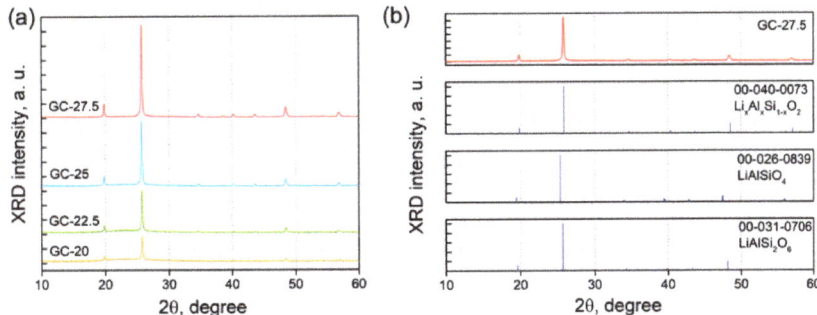

Figure 6. (**a**) XRD patterns of the glass-ceramic samples; (**b**) Standard PDF cards of $Li_xAl_xSi_{1-x}O_2$, $LiAlSiO_4$, and $LiAlSi_2O_6$ for comparison.

Increasing the duration of the holding time at a temperature of 710 °C up to 27.5 h results in a substantial increase in the content of the crystalline phase and a decrease in the fraction of the amorphous phase, which cannot be evidently observed in the XRD patterns. The evaluation of the crystalline phase content and crystallite size is presented in Table 2. The crystallite size remains relatively constant at 33–35 nm regardless of the holding time. However, the degree of crystallization of the samples increases from approximately 14% to 49% after 20 h and 27.5 h of the second stage of the thermal treatment, respectively. It is worth noting that β-eucryptite exhibits a negative CTE coefficient (α), with the average values of α = −6.2 ppm/K in the c-axis direction within the temperature range of 20–1000 °C. Therefore, significant changes in the thermomechanical properties can be expected for the samples subjected to heat treatment at 710 °C within the specified time range. These findings are consistent with the results obtained from the Raman spectroscopy, in which an increase in the holding time leads to a notable enhancement in the intensities of the bands associated with the formation of the crystalline phase (Figure 5).

Table 2. Calculated crystallized fraction, crystallite size, and CTE values in different temperature ranges ($CTE_{T1...T2}$) for the glass-ceramic samples.

Sample	Duration, h	Crystallized Fraction, %	Crystallite Size, nm	$CTE_{0...50\,°C}$, ppm/K	$CTE_{-50...400\,°C}$, ppm/K	$CTE_{20...500\,°C}$, ppm/K
GC-20	20	14	33	3.64	4.34	4.73
GC-22.5	22.5	24	33	3.13	3.88	4.25
GC-25	25	32	35	0.51	0.75	0.94
GC-27.5	27.5	49	35	−0.77	−0.49	−0.34

TEM microscopy was used to investigate the microstructure of the glass-ceramics obtained after the two-stage thermal treatment of glass. The TEM image shown in Figure 7a depicts the microstructure of the GC-27.5 sample, which is typical for LAS glass-ceramics. A comprehensive analysis of the TEM images reveals that the microstructure of the sample consists of crystallites of varying sizes, referred to as small and large particles, which converge with the data regarding the microstructures of other LAS glasses [9,44].

The size of the small particles was determined from a few TEM images and found to be approximately 12 ± 2 nm, while the size of the large particles was calculated to be around 40 ± 11 nm (Figure 7b). The small particles are most likely the nucleation centers for the growth of LAS-phase crystals in the form of large particles. These small particles can be described as nanocrystals of $ZrTiO_4$ and a mixed form of $Zr_{1-x}Ti_{1+x}O_4$, a finding that has been confirmed by numerous researchers [44–47]. It is worth noting that the electron beam used during TEM imaging did have an impact on the glass-ceramic sample. However, the size measurements obtained from the TEM images for the LAS crystals (40 ± 11 nm) coincide well with the estimation (33–35 nm) using the XRD analysis.

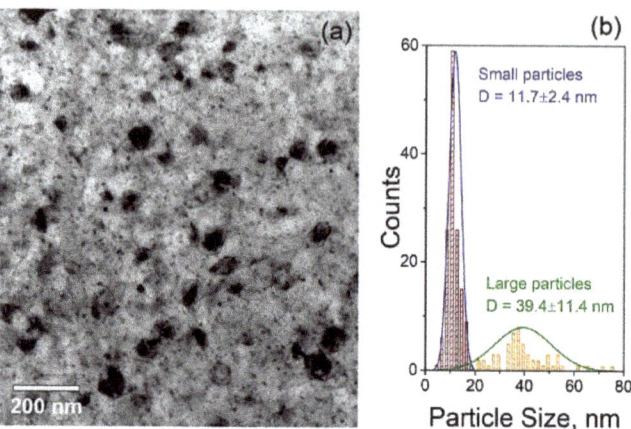

Figure 7. (a) TEM image of the glass-ceramic sample GC-27.5; (b) Particle size distribution as extracted from the TEM image in (a).

Dilatometric curves were recorded over a wide temperature range (−120 to 500 °C) to investigate the effect of the heat treatment duration during the second stage of crystallization on the thermomechanical properties of glass-ceramics. Figure 8 displays the relative elongation curves and the corresponding CTE values as a function of temperature. By varying the holding times at 710 °C, it was possible to smoothly adjust the positions of the relative elongation curves for the samples. The CTE values at 100 °C for the GC-20 and GC-22.5 samples exhibited only minor differences and are equal to 4.0 and 3.5 ppm/K, respectively. The most significant changes in the thermomechanical properties of the heat-treated specimens were observed within a range of holding times between 25 and 27.5 h, as indicated by a transition in the CTE value from positive (0.5 ppm/K) for the GC-25 sample to negative (−0.5 ppm/K) for the GC-27.5 sample. Increasing the holding time up to 30 h resulted in negligible changes in the CTE, with the elongation curve closely resembling the one of the GC-27.5 sample.

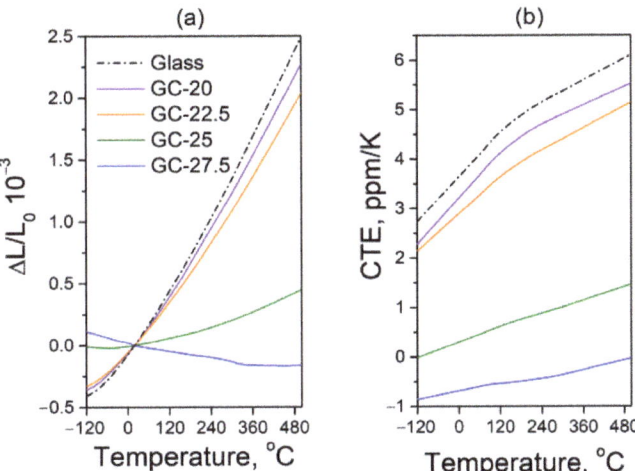

Figure 8. (a) $\Delta L/L_0$ and (b) CTE curves of the bulk glass and glass-ceramic samples.

It can be hypothesized that at a temperature of 710 °C and a holding time of 27.5 h, the primary crystallization processes have already finalized, and prolonging the heat treatment duration does not substantially affect the structure and properties of the LAS glass-ceramics. Thus, by manipulating the duration of the holding time during the temperature treatment of the investigated samples within the confined range of 25 to 27.5 h, it becomes feasible to fine-tune the CTE curve and, moreover, to alter its sign while preserving the stability of the CTE values in close proximity to zero.

The observed variations in the $\Delta L/L_0$ curves as the holding time increases are consistent with the changes observed in the Raman spectra, where an increase in band intensity is registered (Figure 5), as well as with the XRD results showing an increase in the intensity of major peaks (Figure 6). The comparative values of the CTE for different temperature ranges are presented in Table 2. It is demonstrated that an increase in the degree of crystallization leads to a decrease in CTE for all cases. These findings pave the way to finely adjust the CTE values of LAS glass-ceramics by manipulating the holding time during the second stage of the heat treatment.

The optical transmission spectra of the obtained glass-ceramic samples with a thickness of 2 mm are shown in Figure 9. Importantly, all of the samples possess optical transparency, with transmittance (T) values ranging from 70% to 85% in the visible region.

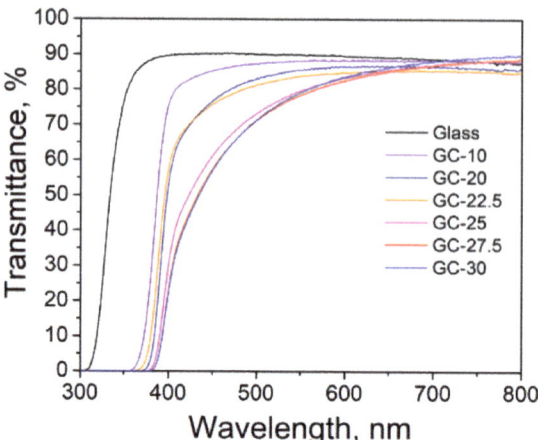

Figure 9. Optical transmittance of 2 mm thick samples before and after crystallization.

4. Conclusions

This study presents a comprehensive investigation into the influence of heat treatment on the structural characteristics of LAS glass-ceramics, which enables the fine-tuning of their thermal expansion coefficient. The temperature–time regime for efficient glass nucleation was determined, and a two-stage heat treatment route was developed to produce transparent glass-ceramics with a relatively high crystalline-phase content. The optimal crystal growth temperature is found to be 710 °C, as this temperature effectively prevents sample cracking caused by internal mechanical stresses during the formation of the crystalline phase. Furthermore, it is demonstrated that the variation in the duration of the heat treatment in the second stage in the range of 20–30 h allows us to precisely control the crystallization of a β-eucryptite-like solid solution as well as the CTE of the glass-ceramics in the range from 0.5 ppm/K to −0.5 ppm/K. The size of crystallites remains relatively unchanged at approximately 33–35 nm, while the degree of crystallization reaches 49%. The results of the study are of interest for researchers for the advancement of integrated optical devices based on temperature-insensitive materials.

Supplementary Materials: The following supporting information can be downloaded at: https://www.mdpi.com/article/10.3390/ceramics7010001/s1, Figure S1: Photo of the synthesized LAS glass block; Figure S2: XRD patterns of one of the glass-ceramic samples and an empty silicon-made low-background specimen holder used to estimate the instrumental background profile; Figure S3: (a) Raman spectra and (b) XRD patterns of initial glass and glass obtained after the nucleation at 670 °C for 2 h; Table S1: Detailed ICP-OES instrumental settings; Table S2: Crystallization peak temperatures determined from the DSC curves (Figures 2 and 3) and calculated $T'_P - T_P$ parameter.

Author Contributions: Conceptualization, A.S.N. and G.Y.S.; methodology, A.S.N. and V.I.S.; investigation, N.V.G., R.O.A., V.I.S., S.S.F. and E.S.I.; writing—original draft preparation, A.S.N. and G.Y.S.; writing—review and editing, A.S.L., N.V.G. and V.N.S.; supervision, V.N.S.; funding acquisition, V.N.S. All authors have read and agreed to the published version of the manuscript.

Funding: This research was funded by the Russian Science Foundation under grant no. 19-19-00613-П.

Institutional Review Board Statement: Not applicable.

Informed Consent Statement: Not applicable.

Data Availability Statement: The data presented in this study are available on request from the corresponding author. The data are not publicly available due to privacy.

Acknowledgments: The authors are grateful to the Center for Collective Use of the Mendeleev University of Chemical Technology for its help with the chemical analysis of the glass.

Conflicts of Interest: The authors declare no conflict of interest.

References

1. Venkateswaran, C.; Sreemoolanadhan, H.; Vaish, R. Lithium aluminosilicate (LAS) glass-ceramics: A review of recent progress. *Int. Mater. Rev.* **2022**, *67*, 620–657. [CrossRef]
2. Dymshits, O.; Shepilov, M.; Zhilin, A. Transparent glass-ceramics for optical applications. *MRS Bull.* **2017**, *42*, 200–205. [CrossRef]
3. Döhring, T.; Jedamzik, R.; Thomas, A.; Hartmann, P. Forty years of ZERODUR mirror substrates for astronomy: Review and outlook. In Proceedings of the SPIE 7018, Advanced Optical and Mechanical Technologies in Telescopes and Instrumentation, Marseille, France, 23–28 June 2008; Volume 7018, pp. 1154–1165. [CrossRef]
4. Semenov, A.P.; Abdulkadyrov, M.A.; Ignatov, A.N.; Patrikeev, V.E.; Pridnya, V.V.; Polyanchikov, A.V.; Sharov, Y.A. Fabrication of blanks, figuring, polishing and testing of segmented astronomic mirrors for SALT and LAMOST projects. In Proceedings of the SPIE 5494, Optical Fabrication, Metrology, and Material Advancements for Telescopes, Glasgow, UK, 21–25 June 2004; Volume 10315, pp. 34–37. [CrossRef]
5. Flügge, J.; Köning, R.; Schötka, E.; Weichert, C.; Köchert, P.; Bosse, H.; Kunzmann, H. Improved measurement performance of the Physikalisch-Technische Bundesanstalt nanometer comparator by integration of a new Zerodur sample carriage. *Opt. Eng.* **2014**, *53*, 122404. [CrossRef]
6. Beverini, N.; Di Virgilio, A.; Belfi, J.; Ortolan, A.; Schreiber, K.U.; Gebauer, A.; Klügel, T. High-accuracy ring laser gyroscopes: Earth rotation rate and relativistic effects. *J. Phys. Conf. Ser.* **2016**, *723*, 012061. [CrossRef]
7. Kuznetsov, A.G.; Molchanov, A.V.; Chirkin, M.V.; Izmailov, E.A. Precise laser gyroscope for autonomous inertial navigation. *Quantum Electron.* **2015**, *45*, 78–88. [CrossRef]
8. Manske, E.; Fröhlich, T.; Füßl, R.; Ortlepp, I.; Mastylo, R.; Blumröder, U.; Dontsov, D.; Kuehnel, M.; Köchert, P. Progress of nanopositioning and nanomeasuring machines for cross-scale measurement with sub-nanometre precision. *Meas. Sci. Technol.* **2020**, *31*, 085005. [CrossRef]
9. Mitra, I. ZERODUR: A glass-ceramic material enabling optical technologies. *Opt. Mater. Express* **2022**, *12*, 3563–3576. [CrossRef]
10. Schulz, H. Thermal expansion of beta eucryptite. *J. Am. Ceram. Soc.* **1974**, *57*, 313–318. [CrossRef]
11. Lichtenstein, A.I.; Jones, R.O.; Xu, H.; Heaney, P.J. Anisotropic thermal expansion in the silicate β-eucryptite: A neutron diffraction and density functional study. *Phys. Rev. B* **1998**, *58*, 6219–6223. [CrossRef]
12. Roy, R.; Agrawal, D.K.; McKinstry, H.A. Very low thermal expansion coefficient materials. *Annu. Rev. Mater. Sci.* **1989**, *19*, 59–81. [CrossRef]
13. Gillery, F.H.; Bush, E.A. Thermal contraction of β-eucryptite ($Li_2O \cdot Al_2O_3 \cdot 2SiO_2$) by X-ray and dilatometer methods. *J. Am. Ceram. Soc.* **1959**, *42*, 175–177. [CrossRef]
14. Guan, J. Femtosecond-laser-written integrated photonics in bulk glass-ceramics Zerodur. *Ceram. Int.* **2021**, *47*, 10189–10192. [CrossRef]
15. Ferreira, P.H.D.; Fabris, D.C.N.; Boas, M.V.; Bezerra, I.G.; Mendonça, C.R.; Zanotto, E.D. Transparent glass-ceramic waveguides made by femtosecond laser writing. *Opt. Laser Technol.* **2021**, *136*, 106742. [CrossRef]
16. Lipatiev, A.; Fedotov, S.; Lotarev, S.; Naumov, A.; Lipateva, T.; Savinkov, V.; Shakhgildyan, G.; Sigaev, V. Direct laser writing of depressed-cladding waveguides in extremely low expansion lithium aluminosilicate glass-ceramics. *Opt. Laser Technol.* **2021**, *138*, 106846. [CrossRef]
17. Khodakovskaya, R.Y. *Chemistry of Titanium-Containing Glasses and Glass Ceramics*; Khimiya: Moscow, Russia, 1978.
18. Khodakovskaya, R.Y.; Sigaev, V.N.; Plutalov, N.F. Phase separation of $Li_2O–Al_2O_3–SiO_2–TiO_2$ glass at the initial stages of crystallization. *Fiz. Khim. Stekla.* **1979**, *5*, 134–140.
19. Sakamoto, A.; Yamamoto, S. Glass-ceramics: Engineering principles and applications. *Int. J. Appl. Glas. Sci.* **2010**, *1*, 237–247. [CrossRef]
20. Bach, H.; Krause, D. (Eds.) *Low Thermal Expansion Glass Ceramics*; Springer: Berlin/Heidelberg, Germany, 2005; pp. 121–235.
21. Liu, X.; Zhou, J.; Zhou, S.; Yue, Y.; Qiu, J. Transparent glass-ceramics functionalized by dispersed crystals. *Prog. Mater. Sci.* **2018**, *97*, 38–96. [CrossRef]
22. Kleebusch, E.; Patzig, C.; Höche, T.; Rüssel, C. A modified B_2O_3 containing $Li_2O-Al_2O_3-SiO_2$ glass with ZrO_2 as nucleating agent-crystallization and microstructure studied by XRD and (S)TEM-EDX. *Ceram. Int.* **2018**, *44*, 19818–19824. [CrossRef]
23. Wu, J.; Lin, C.; Liu, J.; Han, L.; Gui, H.; Li, C.; Liu, T.; Lu, A. The effect of complex nucleating agent on the crystallization, phase formation and performances in lithium aluminum silicate (LAS) glasses. *J. Non-Cryst. Solids* **2019**, *521*, 119486. [CrossRef]
24. Dymshits, O.; Bachina, A.; Alekseeva, I.; Golubkov, V.; Tsenter, M.; Zapalova, S.; Bogdanov, K.; Danilovich, D.; Zhilin, A. Phase transformations upon formation of transparent lithium alumosilicate glass-ceramics nucleated by yttrium niobates. *Ceramics* **2023**, *6*, 1490–1507. [CrossRef]
25. Maier, V.; Muller, G. Mechanism of oxide nucleation in lithium aluminosilicate glass-ceramics. *J. Am. Ceram. Soc.* **1987**, *70*, 176–178. [CrossRef]
26. Venkateswaran, C.; Sharma, S.C.; Pant, B.; Chauhan, V.S.; Vaish, R. Crystallisation studies on site saturated lithium aluminosilicate (LAS) glass. *Thermochim. Acta* **2019**, *679*, 178311. [CrossRef]
27. Kumar, A.; Chakrabarti, A.; Shekhawat, M.S.; Molla, A.R. Transparent ultra-low expansion lithium aluminosilicate glass-ceramics: Crystallization kinetics, structural and optical properties. *Thermochim. Acta* **2019**, *676*, 155–163. [CrossRef]

28. Zhang, R.; Yi, L.; Kong, F.; Liang, X.; Yin, Z.; Rao, Y.; Wang, D.; Chen, Z.; Yu, X.; Jiang, H.; et al. Rapid preparation of low thermal expansion transparent LAS nanocrystalline glass by one-step thermoelectric treatment. *Ceram. Int.* **2021**, *47*, 34380–34387. [CrossRef]
29. Komatsu, T.; Honma, T. Laser patterning and growth mechanism of orientation designed crystals in oxide glasses: A review. *J. Solid State Chem.* **2019**, *275*, 210–222. [CrossRef]
30. Wang, X.; Yu, H.; Li, P.; Zhang, Y.; Wen, Y.; Qiu, Y.; Liu, Z.; Li, Y.; Liu, L. Femtosecond laser-based processing methods and their applications in optical device manufacturing: A review. *Opt. Laser Technol.* **2021**, *135*, 106687. [CrossRef]
31. Zhang, B.; Wang, L.; Chen, F. Recent advances in femtosecond laser processing of $LiNbO_3$ crystals for photonic applications. *Laser Photonics Rev.* **2020**, *14*, 1900407. [CrossRef]
32. Lei, Y.; Wang, H.; Skuja, L.; Kühn, B.; Franz, B.; Svirko, Y.; Kazansky, P.G. Ultrafast laser writing in different types of silica glass. *Laser Photonics Rev.* **2023**, *17*, 2200978. [CrossRef]
33. Stoian, R.; Colombier, J.-P. Advances in ultrafast laser structuring of materials at the nanoscale. *Nanophotonics* **2020**, *9*, 4665–4688. [CrossRef]
34. Khalil, A.A.; Bérubé, J.-P.; Danto, S.; Desmoulin, J.-C.; Cardinal, T.; Petit, Y.; Vallée, R.; Canioni, L. Direct laser writing of a new type of waveguides in silver containing glasses. *Sci. Rep.* **2017**, *7*, 11124. [CrossRef]
35. Shakhgildyan, G.Y.; Alekseev, R.O.; Golubev, N.V.; Savinkov, V.I.; Naumov, A.S.; Presnyakova, N.N.; Sigaev, V.N. One-step crystallization of gahnite glass-ceramics in a wide thermal gradient. *Chemengineering* **2023**, *7*, 37. [CrossRef]
36. Langford, J.I.; Wilson, A.J.C. Scherrer after sixty years: A survey and some new results in the determination of crystallite size. *J. Appl. Crystallogr.* **1978**, *11*, 102–113. [CrossRef]
37. Kleebusch, E.; Patzig, C.; Höche, T.; Rüssel, C. The evidence of phase separation droplets in the crystallization process of a Li_2O-Al_2O_3-SiO_2 glass with TiO_2 as nucleating agent—An X-ray diffraction and (S)TEM-study supported by EDX-analysis. *Ceram. Int.* **2018**, *44*, 2919–2926. [CrossRef]
38. Hartmann, P.; Jedamzik, R.; Carré, A.; Krieg, J.; Westerhoff, T. Glass ceramic ZERODUR®: Even closer to zero thermal expansion: A Review, Part 1. *J. Astron. Telesc. Instrum. Syst.* **2021**, *7*, 020901. [CrossRef]
39. Hartmann, P.; Jedamzik, R.; Carré, A.; Krieg, J.; Westerhoff, T. Glass ceramic ZERODUR®: Even closer to zero thermal expansion: A Review, Part 2. *J. Astron. Telesc. Instrum. Syst.* **2021**, *7*, 020902. [CrossRef]
40. Wurth, R.; Muñoz, F.; Müller, M.; Rüssel, C. Crystal growth in a multicomponent lithia aluminosilicate glass. *Mater. Chem. Phys.* **2009**, *116*, 433–437. [CrossRef]
41. Li, M.; Xiong, C.; Ma, Y.; Jiang, H. Study on crystallization process of Li_2O–Al_2O_3–SiO_2 glass-ceramics based on in situ analysis. *Materials* **2022**, *15*, 8006. [CrossRef]
42. Marotta, A.; Buri, A.; Branda, F. Nucleation in glass and differential thermal analysis. *J. Mater. Sci.* **1981**, *16*, 341–344. [CrossRef]
43. Loshmanov, A.A.; Sigaev, V.N.; Khodakovskaya, R.Y.; Pavlushkin, N.M.; Yamzin, I.I. Small-angle neutron scattering on silica glasses containing titania. *J. Appl. Crystallogr.* **1974**, *7*, 207–210. [CrossRef]
44. Kleebusch, E.; Patzig, C.; Höche, T.; Rüssel, C. Effect of the concentrations of nucleating agents ZrO_2 and TiO_2 on the crystallization of Li_2O–Al_2O_3–SiO_2 glass: An X-ray diffraction and TEM investigation. *J. Mater. Sci.* **2016**, *51*, 10127–10138. [CrossRef]
45. Kleebusch, E.; Thieme, C.; Patzig, C.; Höche, T.; Rüssel, C. Crystallization of lithium aluminosilicate and microstructure of a lithium alumino borosilicate glass designed for zero thermal expansion. *Ceram. Int.* **2023**, *49*, 21246–21254. [CrossRef]
46. Bhattacharyya, S.; Höche, T.; Jinschek, J.R.; Avramov, I.; Wurth, R.; Müller, M.; Rüssel, C. Direct evidence of Al-rich layers around nanosized $ZrTiO_4$ in glass: Putting the role of nucleation agents in perspective. *Cryst. Growth Des.* **2010**, *10*, 379–385. [CrossRef]
47. Höche, T.; Patzig, C.; Gemming, T.; Wurth, R.; Rüssel, C.; Avramov, I. Temporal evolution of diffusion barriers surrounding $ZrTiO_4$ nuclei in lithia aluminosilicate glass-ceramics. *Cryst. Growth Des.* **2012**, *12*, 1556–1563. [CrossRef]
48. Höche, T.; Mäder, M.; Bhattacharyya, S.; Henderson, G.S.; Gemming, T.; Wurth, R.; Rüssel, C.; Avramov, I. $ZrTiO_4$ crystallisation in nanosized liquid–liquid phase-separation droplets in glass—A quantitative XANES study. *CrystEngComm* **2011**, *13*, 2550–2556. [CrossRef]
49. Serbena, F.C.; Zanotto, E.D. Internal residual stresses in glass-ceramics: A review. *J. Non-Cryst. Solids* **2012**, *358*, 975–984. [CrossRef]
50. Serbena, F.C.; Soares, V.O.; Peitl, O.; Pinto, H.; Muccillo, R.; Zanotto, E.D. Internal residual Stresses in sintered and commercial low expansion Li_2O-Al_2O_3-SiO_2 glass-ceramics. *J. Am. Ceram. Soc.* **2011**, *94*, 1206–1214. [CrossRef]
51. Ross, S.; Welsch, A.-M.; Behrens, H. Lithium conductivity in glasses of the Li_2O–Al_2O_3–SiO_2 system. *Phys. Chem. Chem. Phys.* **2015**, *17*, 465–474. [CrossRef]
52. Wei, A.; Liu, Z.; Zhang, F.; Ma, M.; Chen, G.; Li, Y. Thermal expansion coefficient tailoring of LAS glass-ceramic for anodic bondable low temperature co-fired ceramic application. *Ceram. Int.* **2020**, *46*, 4771–4777. [CrossRef]

Disclaimer/Publisher's Note: The statements, opinions and data contained in all publications are solely those of the individual author(s) and contributor(s) and not of MDPI and/or the editor(s). MDPI and/or the editor(s) disclaim responsibility for any injury to people or property resulting from any ideas, methods, instructions or products referred to in the content.

Influence of Alkali Metal Ions on the Structural and Spectroscopic Properties of Sm^{3+}-Doped Silicate Glasses

Israel R. Montoya Matos

Department of Civil Engineering, Universidad de Lima, Av. Javier Prado Este 4600, Lima 15023, Peru; imontoya@ulima.edu.pe

Abstract: In the present work, the influence of alkali ions (Li, Na, K) on the structural and spectroscopic properties of silica glasses doped with Sm^{3+} was investigated. Infrared and Raman spectroscopy techniques were used to investigate the structural properties of the alkali silicate glasses. The optical absorption showed bands characteristic of Sm^{3+} ions in alkali silicate glasses, and this was investigated. The Judd–Ofelt theory was applied to evaluate the phenomenological intensity parameters (Ω_2, Ω_4, and Ω_6) of the optical absorption measurements. The multi-channel visible and near infrared emission transitions originating from the $^4G_{5/2}$-emitting state of the Sm^{3+} in alkali silicate glasses with a maximum phonon energy of ~1050 cm^{-1} were investigated. From the evaluated Judd–Ofelt parameters, radiative parameters such as spontaneous emission probabilities, radiative lifetimes, branching ratios, and stimulated emission cross-sections were calculated. The recorded luminescence spectra regions revealed intense green, orange, red, and near-infrared emission bands, providing new traces for developing tunable laser and optoelectronic devices.

Keywords: alkali metal; structural and spectroscopic properties; Judd–Ofelt parameters

1. Introduction

Due to their potential for use as lasers and phosphors in a variety of hosts, including glasses, crystals, and transparent vitro-ceramics that exhibit luminescent transitions in the VIS and NIR spectra regions, Ln^{3+}-doped ions have attracted considerable technological interest [1]. According to research, the optical spectra of the rare earth ions can be used as structural probes to determine the local field parameters within a specific host glass because the energy levels, profiles, and intensities of the absorption and emission bands depend on how charges are distributed in the first coordination shells of the rare earth ions [2,3]. In oxide glasses, silicate glasses are one of the most popular glass hosts for making optical fiber lasers and amplifiers. Several papers have been published on the optical properties of rare earth ions in different glasses [4–6], but only a few of them have been concerned with Sm^{3+} [7–10]. The significant energy gap (less non-radiative decay) between the $^4G_{5/2}$ level and the next lower-lying energy level, $^6F_{11/2}$, which is roughly 7200 cm^{-1}, causes glasses doped with Sm^{3+} ions to have relatively high quantum efficiencies. It has been known for a long time that Sm^{3+} ion provide very strong luminescence in the orange and red spectral regions in a variety of lattices [11,12]. However, only a few attempts have been made to explore the possibility of using the orange-red luminescence of Sm^{3+} ions for the development of visible optical devices. The main reason for not carrying out spectral studies on Sm^{3+} ions doped in glasses is the complicated structure of the 4f^6 configuration of this ion because these glasses show different channels that lead to luminescence-quenching and have strong phosphorescent intensities [13].

In the present work, the spectroscopic properties of Sm^{3+} ions were used to investigate the local structures of alkali (i.e., Li, Na, and K) oxides in silica. It has been known for a long time that the gradual replacement of one alkali oxide by another induces nonlinear changes in certain physical properties of glasses [14,15], and the explanation for this effect in terms of the atomic structure is not simply due to the amorphous nature of the glass.

2. Materials and Methods

Silicate glasses have the following composition: 33.0 R_2O + 66.0 SiO_2 + 1.0 Sm_2O_3 (mol%), where R = Li, Na, and K. The glasses were prepared using the traditional melting quenching technique. The laboratory chemicals used included purified sand Li_2CO_3, Na_2CO_3, K_2CO_3, SiO_2, and Sm_2O_3 (99.99% purity grade from Sigma-Aldrich (St. Louis, MO, USA)), and these were used for sample preparation. Since carbonates and silica are highly hygroscopic, they were dried at 200 °C for 1–2 h. The mixture was placed in a platinum crucible and heated at 700 to 800 °C to eliminate CO_2 [16], and then it was melted at 1550 °C for 2 h in air. The melted mixture was quickly poured into a preheated stainless steel mold and annealed at 350 °C for 4 h. After that, it was cooled down slowly to room temperature at a rate of 5 °C/min. Table 1 shows the compositions, appearances, and melting temperatures of the prepared samples.

Table 1. Chemical compositions of the glass samples.

Sample Notation	Glass Compositions (mol%)	Glass Appearances	Melting Temperatures (°C)
LS-Sm	33.0 Li_2O + 66.0 SiO_2 + 1.0 Sm_2O_3	Transparent glass	1550
NS-Sm	33.0 Na_2O + 66.0 SiO_2 + 1.0 Sm_2O_3	Transparent glass	1550
KS-Sm	33.0 K_2O + 66.0 SiO_2 + 1.0 Sm_2O_3	Transparent glass	1550

The densities of the prepared samples were measured using the Archimedes method (with an analytical balance Shimadzu AUW220D, 0.1 mg/0.01 mg) using distilled water as the immersion liquid. The refractive index was measured using a SOPRA GES-5E ellipsometer. The FTIR transmittance spectra were measured using a Bomem MB100 spectrometer using KBr pellets. The Raman spectra were measured using an HeNe laser as the excitation source. The absorption spectra in the UV-Vis-Nir region were recorded using a high-performance spectrometer (PerkinElmer model LAMBDA 1050) with a spectral resolution of 0.2 nm. The photoluminescence (PL) data were measured using a CryLas GmbH 488 nm CW laser as a pumping source. The PL signal was dispersed by an Acton SP 2300 monochromator and detected by a Pixis 256E CCD. The detected signal was fed to an SR 430 multichannel analyzer and transferred to a computer running acquisition software (OriginPro).

3. Results and Discussions

3.1. Structural Analysis

One of the most successful techniques for structural analysis is infrared spectroscopy, which makes it possible to identify structural features related to both the anionic sites containing the modifying cations and the local units that form the glass network [17]. Figure 1 shows the vibrational spectra of the alkali silicate glasses, which exhibit prominent bands at ~1037 cm^{-1} for LS, ~1060 cm^{-1} for NS, and ~1013 cm^{-1} for KS due to Si–O–Si stretching within the tetrahedral by the presence of alkali ions. These bands are sharpened by the addition of rare earth ions. The introduction of rare earth ions in the matrix will alter the environment of the defect centers in the silica [18]. Also, the relative intensities of the peaks corresponding to the Si–O stretching from the tetrahedral and the non-bridged Si–O were nearly the same for the LS and NS samples, but for the KS sample, it was possible to observe a shift in the band at ~1013 cm^{-1} to lower wavenumbers, suggesting that the effect of K on the Si–O stretching within the tetrahedral was slightly different from that of Li and Na. The bands at ~916 cm^{-1} for LS, ~891 cm^{-1} for NS, and 884 cm^{-1} for KS were due to the stretching of the non-bridged terminal Si–O. The addition of alkali ions would decrease the local symmetry due to the formation of non-bridging oxygen bonds, giving rise to a stretching mode at ~900 cm^{-1} [19]. Comparing the spectra shown in Figure 1 and described in Table 2, all of them presented shoulders at ~900 cm^{-1}, though they were more pronounced for LS and NS samples, indicating the formation of non-bridged terminal Si–O.

The bands at ~774 cm^{-1} for LS, ~752 cm^{-1} for NS, and ~745 cm^{-1} for KS were due to the stretching of the bridged Si–O–Si symmetric vibrations; more precisely, they were due to the bending mode of the bridging oxygen perpendicular to the Si–Si axis within the Si–O–Si plane and were at ~470 cm^{-1} for LS and NS and ~450 cm^{-1} for KS due to the Si–O rocking motion.

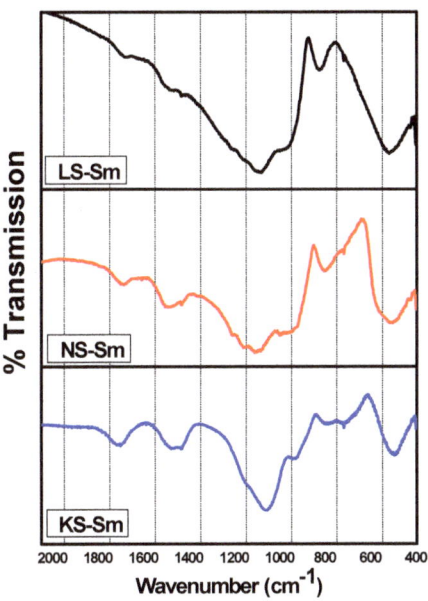

Figure 1. FTIR absorption spectra of the different alkali silicate glasses.

Table 2. Vibration mode FTIR spectra of the alkali silicate glasses.

Vibration Mode	LS-Sm Wavenumber (cm^{-1})	NS-Sm Wavenumber (cm^{-1})	KS-Sm Wavenumber (cm^{-1})
Si–O–Si bending vibration	472	468	447
O–Si–O bending vibration	774	752	745
Si–O–Si symmetric stretching	916	891	884
Si–O–Si asymmetric stretching	1037	1060	1013

Figure 2 shows the Raman spectra of the samples. These spectra consisted of three intense peaks as representative cases, and the band assignments are presented in Table 3. Prominent features of the silica glass Raman spectrum were shown in Matson et al. [20]. The Raman spectrum of silica contains a large and asymmetric band near ~440 cm^{-1} and a sharp band at ~492 cm^{-1} related to the SiO$_4$ tetrahedra. A broad band at ~800 cm^{-1} is related to the network of the SiO$_2$ glass, a band at ~600 cm^{-1} is related to defects, bands at 1060 and 1190 cm^{-1} are related to Si–O–Si vibrations, and a band at ~950 cm^{-1} is related to crystalline metasilicates containing chains of SiO$_4$ tetrahedra. According to Matson et al. [20], the introduction of alkali ions in the silica network induces changes in the Raman spectrum depending on the ion type and concentration. The stretching of non-bonding oxygen on SiO$_4$ tetrahedra due to the alkali ion appeared at ~1100 cm^{-1} [20]. The results shown in Figure 2 showed that intense bands were observed for all samples at close to 600 cm^{-1}, and they were related to defects in the silica network and were likely induced by the alkali ions (the creation of non-bonding oxygen atoms). Consistently, the bands close to ~1100 cm^{-1} and ~950 cm^{-1} were also observed for all samples while the band at ~800 cm^{-1}, related to the silica network, was not observed. This suggested that the incorporation of the rare earth oxides studied in the alkali silicate glasses caused more non-bridging oxygen to occur, and consequentially, it led to more polymerization in the network. It is expected that the presence of rare earth oxides as modifiers contribute three NBO (non-bridging-oxygen) atoms to a glass system while traditional modifiers (Li_2O, Na_2O, and K_2O) contribute only one NBO (non-bridging-oxygen) atom [21].

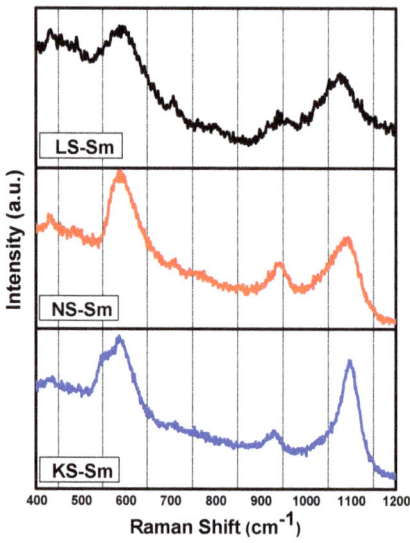

Figure 2. Raman spectra of the different alkali silicate glasses.

Table 3. Vibration-mode Raman spectra of the alkali silicate glasses.

Vibration mode	LS-Sm Wavenumber (cm^{-1})	NS-Sm Wavenumber (cm^{-1})	KS-Sm Wavenumber (cm^{-1})
Si–O	591	589	587
Si–O–Si symmetric stretching	939	942	930
Si–O–Si antisymmetric stretching	1073	1094	1097

Physical parameters such as the densities, refractive indexes, concentrations of Sm^{3+} ions, polaron radii, inter-ionic distances, field strengths, reflection losses, optical dielectrics, and dielectric constants of the glasses were calculated using standard equations [22,23] and are presented in Table 4.

3.2. Optical Properties

The absorption bands of the RE ions in a host glass can be expressed as oscillator strengths, and the electronic transitions can be calculated using the expression:

$$f_{exp} = 4.318 \times 10^{-9} \int \varepsilon(\nu) d\nu \qquad (1)$$

where the integral represents the area under the absorption curve and $\varepsilon(\nu)$ is the molar extinction coefficient. According to the Judd–Ofelt (JO) theory [24,25], which defines a set of three parameters (Ω_2, Ω_4, and Ω_6) susceptible to the environment of the RE ion, the parameter Ω_2 is related to the covalence of the metal-ligand bond while Ω_4 and Ω_6 are related to the rigidity of the host matrix. In the Judd–Ofelt theory, the oscillator strength,

f_{cal}, of the spectral intensity of an electric dipole absorption transition from the initial state, aJ, to the final state, bJ',

$$f_{cal}(aJ, bJ') = \frac{8\pi^2 mcv}{3h(2J+1)} \left[\frac{(n^2+2)^2}{9n}\right] x \sum_{\lambda=2,4,6} \Omega_\lambda \left\langle aJ \left| U^\lambda \right| bJ' \right\rangle^2 \quad (2)$$

where m is the mass of an electron, c is the velocity of the light, h is the Planck's constant, v is the wavenumber of the absorption peak, n is the refractive index, and U^λ is the doubly reduced matrix element.

Table 4. Physical properties of the Sm^{3+}-doped alkali metal ion silicate samples.

Physical Properties	LS-Sm	NS-Sm	KS-Sm
Density, ρ (gm/cm^3) ($\pm$ 0.0005)	2.4552	2.5396	2.5598
Refractive index (n) (585 nm)	1.4930	1.4490	1.4310
Concentration of Sm^{3+}, N ($\times 10^{20}$ ions/cm^3)	2.1896	1.8854	1.7370
Molar volume (V_m) (cm^3/mol)	27.5024	31.9404	34.6697
Polaron radius, r_p (Å)	6.6846	7.0264	7.2211
Interionic distance, r_i (nm)	1.6591	1.7439	1.7923
Field strength, F ($\times 10^{16}$ cm^{-2})	6.7139	6.0766	5.7533
Reflection losses (R_L) ($\times 10^{-2}$)	3.9107	3.3614	3.1433
Molar refraction (R_m) (cm^3/mol)	7.9928	8.5671	8.9742
Optical dielectric constant (P$\partial t/\partial P$)	1.2290	1.0996	1.0478
Dielectric constant (ε)	2.2290	2.0996	2.0478
Molar polarizability (α_m) $\times 10^{-22}$ cm^3	3.1687	3.3964	3.5578

The optical absorption spectra of the Sm^{3+}-doped different alkali silicate glasses (Li, Na, and K) recorded at temperature room are shown in Figure 3. The spectral behavior of the Sm^{3+} in the glasses was very similar to that found in fluorophosphate [26], borosulphate [27], and germanate [28] glasses. Seven absorption bands were shown for each sample, as described in Table 5, which were associated with the absorption transitions from the $^6H_{5/2}$ ground state to the excited states, as labeled in Figure 3. The observed band positions of different absorption peaks of the samples were due to the interactions between the alkali ions and the crystal field. The band shapes of the ions in all three alkali silicate glasses were similar, with small differences in the half-band widths and peak positions. These results suggested that the Sm^{3+} ions resided in essentially the same sites as the three alkali ions. The calculated $\bar{\beta}$ and δ values of the glasses are presented in Table 5. The observed the nephelauxetic effect [29,30] was increased with the increase in the alkali ions as follows: K > Na > Li. The negative results demonstrated that the Sm^{3+} ions and their surrounding ligands were primarily ionic in character. Sm^{3+} has a $4f^5$ electron configuration which is characterized by $^{2S+1}L_J$ free-ion levels. With knowledge of the Sm^{3+} concentrations, sample thicknesses, peaks positions, and peak areas, the experimental oscillator strengths were obtained by using Equation (1). From the experimentally measured oscillator strengths and doubly reduced matrix elements [31], the J–O intensity parameters were obtained using Equation (2). The measured, as well calculated, oscillator strengths and J–O intensity parameters of the Sm^{3+} in the alkali silicate glasses are presented in Tables 6 and 7, respectively. The magnitudes of J–O intensity parameters are important for investigations of glass structures and transition properties of rare earth ions. In general, the Ω_2 parameter is an indicator of the covalence of a metal–ligand bond (short-range effect) and the Ω_4 and Ω_6 parameters provide information about the rigidity of a host glass's matrix (long-range effects) [32].

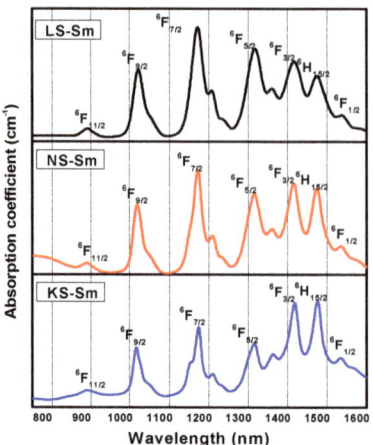

Figure 3. Optical absorption spectra of the Sm^{3+}-doped alkali silicate glasses recorded at room temperature in the NIR region.

Table 5. Bands positions (cm^{-1}) and their assignments for the Sm^{3+}-doped alkali silicates.

Transitions $^6H_{5/2}$	LS-Sm	NS-Sm	KS-Sm
$^6F_{11/2}$	10,626.99	10,655.73	10,672.36
$^6F_{9/2}$	9325.81	9353.91	9369.21
$^6F_{7/2}$	8185.95	8183.33	8169.19
$^6F_{5/2}$	7316.94	7325.95	7322.03
$^6F_{3/2}$	6815.18	6821.16	6806.60
$^6H_{15/2}$	6558.67	6560.53	6549.88
$^6F_{1/2}$	6288.49	6306.04	6309.73
$\bar{\beta}$	1.01436	1.01569	1.01573
δ	−0.01416	−0.01544	−0.01548

Table 6. Experimental (f_{exp}) and calculated (f_{cal}) oscillator strengths ($\times 10^{-6}$) and rms (σ_{rms}) for the Sm^{3+} alkali silicate glasses.

Transitions $^6H_{5/2}$	LS-Sm		NS-Sm		KS-Sm	
	f_{exp}	f_{cal}	f_{exp}	f_{cal}	f_{exp}	f_{cal}
$^6F_{11/2}$	0.602	0.568	0.363	0.407	0.309	0.205
$^6F_{9/2}$	4.720	3.440	3.390	2.470	1.640	1.230
$^6F_{7/2}$	3.490	4.660	2.490	3.33	1.220	1.610
$^6F_{5/2}$	2.360	1.890	1.630	1.370	0.648	0.595
$^6F_{3/2}$	1.210	1.060	1.170	0.874	0.760	0.475
$^6H_{15/2} + {}^6F_{1/2}$	0.897	0.326	1.091	0.383	0.806	0.307
σ_{rms}	0.9856		0.8310		0.5036	

In the values for Ω_2, Ω_4, and Ω_6 in Table 7 for the Sm^{3+}-doped alkali silicate glasses, we observed variations in the intensities of the parameters as follows: $\Omega_6 > \Omega_4 > \Omega_2$. The lithium silicate glass matrix exhibited higher Ω_2 values and the potassium silicate glass matrix was minimal, indicating lower covalence values for these compositions. It has been proposed that in oxide glasses, a rare earth ion is surrounded by eight neighboring oxygen atoms belonging to the corners of SiO_4, forming a tetrahedral. Normally, the intensity parameter Ω_2, which indicates covalence, decreases with decreases in the intensity of the hypersensitive transition.

The hypersensitive transitions for the Sm^{3+} ions were $^6H_{5/2} \rightarrow {}^6F_{1/2}$ and $^6F_{3/2}$. The hypersensitivity of a transition is proportional to the nephelauxetic ratio, indicating the RE–O bond's ionic nature [33]. The Ω_6 parameter was the largest in the glasses, showing that the Sm–O bond was more ionic in these glasses than in other glasses and indicating the higher rigidity of the glasses. The shift in the peak wavelengths of the hypersensitive transitions towards shorter wavelengths increased with the size of the alkali ions in the glass matrix, indicating decreases in the ionic nature of the RE–O bonds. Also, these changes were not the result of a large-scale structural rearrangement of the local glass network, but rather, they were primarily caused by the interactions between the alkali ions and the matrix.

Table 7. A comparison of the Judd–Ofelt parameters (Ω_λ) ($\times 10^{-20}$ cm^2) in different hosts.

Host Matrix	Ω_2	Ω_4	Ω_6	$\chi = \Omega_4/\Omega_6$	Order
LS-Sm^{3+} (present work)	1.28	3.14	4.53	0.69	$\Omega_6 > \Omega_4 > \Omega_2$
LCN borate [34]	0.84	4.00	5.02	0.79	$\Omega_6 > \Omega_4 > \Omega_2$
30Li$_2$O:70B$_2$O$_3$-Pr^{3+} [35]	0.10	4.71	5.28	0.89	$\Omega_6 > \Omega_4 > \Omega_2$
NS-Sm^{3+} (present work)	1.27	2.69	3.15	0.85	$\Omega_6 > \Omega_4 > \Omega_2$
NaZnBS-Sm^{3+} [27]	0.55	9.68	9.77	0.99	$\Omega_6 > \Omega_4 > \Omega_2$
30Na$_2$O:70B$_2$O$_3$-Pr^{3+} [35]	0.98	4.76	4.86	0.97	$\Omega_6 > \Omega_4 > \Omega_2$
KS-Sm^{3+} (present work)	1.04	1.12	1.61	0.69	$\Omega_6 > \Omega_4 > \Omega_2$
KZnBS-Sm^{3+} [27]	0.18	11.37	11.45	0.99	$\Omega_6 > \Omega_4 > \Omega_2$
LKG [11]	0.63	4.05	4.69	0.86	$\Omega_6 > \Omega_4 > \Omega_2$

3.3. Emission Spectra and Radiative Properties

The emission spectra under the 488 nm excitation wavelength consisted of potential green, yellow, and orange-reddish emissions at ~565 nm ($^4G_{5/2} \rightarrow {}^6H_{5/2}$ magnetic dipole (MD) transition), ~600 nm ($^4G_{5/2} \rightarrow {}^6H_{7/2}$ mixed magnetic-electric dipole transition), ~650 nm ($^4G_{5/2} \rightarrow {}^6H_{9/2}$ electric dipole transition), and ~710 nm ($^4G_{5/2} \rightarrow {}^6H_{11/2}$ electronic dipole transition), all of which were recorded at room temperature in the VIS regions, as shown in Figure 4. The peaks in the NIR region centered at ~915 nm ($^4G_{5/2} \rightarrow {}^6F_{3/2}$), ~960 nm ($^4G_{5/2} \rightarrow {}^6F_{5/2}$), ~1045 nm ($^4G_{5/2} \rightarrow {}^6F_{7/2}$), ~1200 nm ($^4G_{5/2} \rightarrow {}^6F_{9/2}$), and ~1300 nm ($^4G_{5/2} \rightarrow {}^6F_{11/2}$) are presented in Figure 5. It is important to note the largest number of transitions observed, including the one at 1300 nm, indicated that the alkali silicate glasses were very efficient systems.

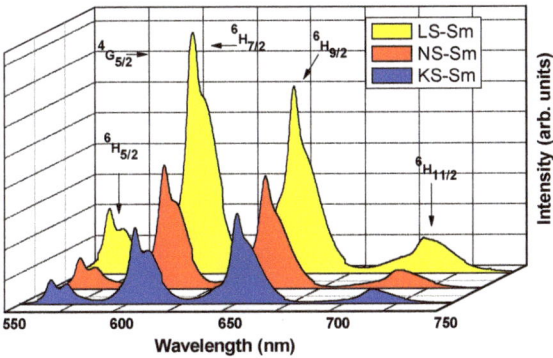

Figure 4. VIS region emission spectra for the alkali silicate glasses under excitation at 488 nm.

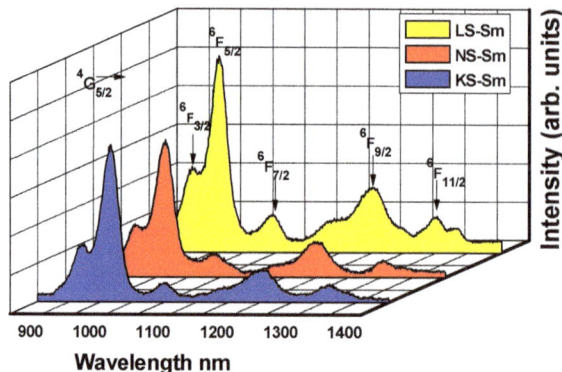

Figure 5. NIR region emission spectra for the alkali silicate glasses under excitation at 488 nm.

The radiative transition probability A_{rad} (aJ and bJ') is given by the expression:

$$A_{rad}(aJ,bJ') = \frac{64\pi^4 v^3 e^2}{3h(2J+1)} \left[\frac{n(n^2+2)^2}{9} S_{ed}\right] \quad (3)$$

where S_{ed} and S_{md} are electric and magnetic dipole line-strengths, respectively. From these values, it is possible to calculate the branching ratio and radiative lifetime of an emitting state, aJ, which is calculated from:

$$\beta_{rad}(aJ,bJ') = \frac{A_{rad}(aJ,bJ')}{\sum_{bJ'} A_{rad}(aJ,bJ')} \quad (4)$$

and

$$\frac{1}{\tau_{rad}} = \sum_{bJ'} A_{rad}(aJ,bJ') \quad (5)$$

In addition, another important radiative parameter property, the fluorescent integrated emission cross-section (σ_p) for stimulated emissions, is estimated from the following equation:

$$\sigma_p(aJ,bJ') = \frac{\lambda_p^4}{8\pi c n^2 \Delta\lambda_{eff}} A_{rad}(aJ,bJ') \quad (6)$$

Due to the characteristic lack of long-range order in glasses, the local environment of RE ions is expected to be slightly different from one site to another, resulting in broad absorption bands. The crystal field splitting is also responsible for the broadening of these bands [26]. It has been assumed that in oxide glasses, an RE ion has eight neighboring oxygen atoms that are shared with the corners of the glass, forming a tetrahedral. In lanthanides, there are certain f–f transitions that are exceptionally sensitive to the local environment, and they are known as hypersensitive transitions ($\Delta J \leq 2$, $\Delta L \leq 2$, and $\Delta S = 0$). For Sm^{+3}, they corresponded to the $^6H_{5/2} \rightarrow {}^6F_{3/2}$ and $^6F_{1/2}$ transitions, which were in the visible range. The addition of alkali ions as network modifiers in oxide glasses induced the formation of non-bridging oxygen atoms, changing the forming cation coordination number. This effect induced changes in the lanthanide–oxygen distances and, therefore, should affect the optical properties of the RE ion in the oxide glass [26].

Using the phenomenological intensity parameters, one can estimate the radiative transition probability values for $A_{rad}(s^{-1})$ (Equation (3)). From these values, it is possible to calculate the branching ratio β_R (Equation (4)) and radiative lifetime τ_{rad} (ms) (Equation (5)) from the $^4G_{5/2}$ excited level to all its lower-lying levels for alkali silicate glasses, as shown in Table 8. The stimulated emission cross-section (Equation (6)) is an important parameter for predicting a better laser performance and high quantum efficiency of glass matrices. Table 9 reports the emission band positions, the effective bandwidths, and the values of emission cross-section $\sigma(\lambda_p)$ for some of the transitions originating from the $^4G_{5/2}$ level of the Sm^{3+} ions in the silicate glasses. Even though the magnitudes of the intensity parameters obtained in the present work were low, the measured $\sigma(\lambda_p)$ values for the $^4G_{5/2} \to {}^6H_{7/2,\,9/2}$ transitions were relatively higher than those reported for other glasses [36–39]. The highest values were observed for the $^4G_{5/2}$-$^6H_{7/2}$ transitions (6.14 × 10^{-22} cm^2, 4.97 × 10^{-22} cm^2, and 2.31 × 10^{-22} cm^2 for the LS, NS, and KS samples, respectively). Despite the values of $\sigma(\lambda_p)$ for the NIR emissions being lower than those of the VIS emissions, they were of the same order of magnitude. Therefore, the novel NIR emissions of the alkali silicate glasses provided a new clue for developing potential NIR optoelectronic devices.

Table 8. Radiative transition probability values (A_{rad} (s^{-1})), branching ratio values (β_R), and radiative lifetime values (τ_{rad} (ms)) for the Sm^{3+}-doped alkali silicate glasses.

Transitions from $^4G_{5/2}$ to	LS-Sm				NS-Sm				KS-Sm			
	ΔE(cm^{-1})	A_{rad}(s^{-1})	β_R	τ_{rad}	ΔE(cm^{-1})	A_{rad}(s^{-1})	β_R	τ_{rad}	ΔE(cm^{-1})	A_{rad}(s^{-1})	β_R	τ_{rad}
$^6H_{5/2}$	17,740	45.40	0.1812	3.99	17,731	36.28	0.1951	5.37	17,728	17.19	0.1906	11.08
$^6H_{7/2}$	16,608	110.618	0.4415		16,584	76.318	0.4104		16,568	34.372	0.3808	
$^6H_{9/2}$	15,407	58.314	0.2327		15,408	46.000	0.2473		15,369	25.510	0.2827	
$^6H_{11/2}$	14,101	26.256	0.1048		14,065	18.945	0.1019		14,025	8.074	0.0895	
$^6F_{3/2}$	10,890	0.800	0.0032		10,905	0.705	0.0038		10,834	0.502	0.0056	
$^6F_{5/2}$	10,424	5.925	0.0236		10,395	5.257	0.0283		10,373	3.405	0.0377	
$^6F_{7/2}$	9578	1.965	0.0078		9606	1.505	0.0081		9533	0.606	0.0067	
$^6F_{9/2}$	8324	0.860	0.0034		8326	0.701	0.0038		8321	0.458	0.0051	
$^6F_{11/2}$	7686	0.419	0.0017		7669	0.271	0.0015		7645	0.129	0.0014	

Table 9. Stimulated emission cross-sections ($\sigma(\lambda_p) \times 10^{-22}$ cm^2) for the emission peak wavelengths (λ_p) in the infrared with the effective line width values ($\Delta\lambda_{eff}$) for the Sm^{3+}-doped alkali silicate glasses.

Transitions from $^4G_{5/2}$ to	LS-Sm			NS-Sm			KS-Sm		
	λ_p(nm)	$\Delta\lambda_{eff}$(nm)	$\sigma(\lambda_p)$	λ_p(nm)	$\Delta\lambda_{eff}$(nm)	$\sigma(\lambda_p)$	λ_p(nm)	$\Delta\lambda_{eff}$(nm)	$\sigma(\lambda_p)$
$^6H_{5/2}$	564	10.14	2.69	564	11.01	2.11	565	12.28	0.92
$^6H_{7/2}$	602	14.16	6.14	603	12.82	4.97	604	12.68	2.31
$^6H_{9/2}$	649	14.76	4.17	650	14.83	3.49	651	13.73	2.16
$^6H_{11/2}$	710	22.91	1.73	711	21.63	1.42	713	18.75	0.72
$^6F_{3/2}$	919	19.03	0.18	919	17.74	0.18	921	19.60	0.12
$^6F_{5/2}$	959	27.37	1.09	962	28.36	1.01	964	22.92	0.83
$^6F_{7/2}$	1044	27.25	0.51	1043	32.19	0.35	1052	24.68	0.19
$^6F_{9/2}$	1199	70.61	0.15	1201	60.99	0.15	1204	61.82	0.10
$^6F_{11/2}$	1301	39.59	0.18	1309	33.35	0.15	1307	41.09	0.06

4. Conclusions

The present work detailed the influence of alkali ions on the structural and spectroscopic properties of Sm^{3+} ion-doped silicon glasses. The FTIR spectra of the doped silica showed that the Si–O–Si bending modes were sharpened with the introduction of rare earth ions. The addition of alkali ions decreased the local symmetry due to the formation of non-bridging oxygen bonds, giving rise to a stretching mode at ~900 cm^{-1}. The optical properties of the Sm^{3+} alkali silicates were measured and analyzed using the Judd–Ofelt theory. The Ω_6 parameter was the largest in the glasses, showing that the Sm–O bonds were more ionic in these glasses than in other glasses and indicating the higher rigidity of these glasses. The intensities of the emission bands in the VIS and NIR regions of the samples containing Li were relatively high compared to those of the samples containing Na and K. The emission spectra showed potential for green, yellow, and orange-reddish emissions. The branching ratios were larger for the $^4G_{5/2} \rightarrow {}^6H_{7/2}$ (VIS region) transition and decreased from the Li and Na to the K silicates, and the ratios for the $^4G_{5/2} \rightarrow {}^6F_{5/2}$ (NIR region) transition decreased as follows: Li > Na > K. The stimulated emission cross-sections obtained were similar to the values reported in the literature for Sm^{3+} in different glasses.

Funding: The author is grateful for the financial support of the Brazilian agencies CAPES, CNPq, and FAPERGS.

Institutional Review Board Statement: Not applicable.

Informed Consent Statement: Not applicable.

Data Availability Statement: Not applicable.

Conflicts of Interest: The author declares no conflict of interest.

References

1. Goodwin, D.W. Spectra and Energy Levels of Rare Earth Ions in Crystals. *Phys. Bull.* **1969**, *20*, 525. [CrossRef]
2. Gatterer, K.; Pucker, G.; Jantscher, W.; Fritzer, H.P.; Arafa, A. Suitability of Nd (III) absorption spectroscopic to probe the structure of glasses from the ternary system Na_2O-B_2O_3-SiO_2. *J. Non-Cryst. Solids* **1998**, *231*, 189–199. [CrossRef]
3. Sailaja, B.; Joyce Stella, R.; Thitumala Rao, G.; Jaya Raja, B.; Pushpa Manjari, V.; Ravikumar, R.V.S.S.N. Physical, structural and spectroscopic investigations of Sm^{3+} doped ZnO mixed alkali borate glass. *J. Mol. Struct.* **2015**, *1096*, 129–235. [CrossRef]
4. Cases, R.; Chamarro, M.A. Judd-Ofelt analysis and multiphonon relaxations of rare earth ions in fluorohafnate glasses. *J. Solid State Chem.* **1991**, *90*, 313. [CrossRef]
5. Basiev, T.; Dergachev, A.Y.; Orlavskii, Y.V.; Prohkorov, A.M. Multiphonon nonradiative relaxation from high-lying levels of Nd^{3+} ions in flouride and oxide laser materials. *J. Lumin.* **1992**, *53*, 19. [CrossRef]
6. Broer, M.M.; Bruce, A.J.; Grodkiewicz, W.H. Resonantly induced refractive index changes in Eu^{3+}- and Er^{3+}-doped silicate and phosphate glasses. *J. Lumin.* **1992**, *53*, 15. [CrossRef]
7. Reisfeld, R.; Bornstein, A.; Boehm, L. Optical characteristics and intensity parameters of Sm^{3+} in GeO_2, ternary germanate, and borate glasses. *J. Solid State Chem.* **1975**, *14*, 14. [CrossRef]
8. Rodriguez, V.D.; Martin, I.R.; Alcala, R.; Cases, R. Optical properties and cross relaxation among Sm^{3+} ions in fluorzincate glasses. *J. Lumin.* **1992**, *54*, 231. [CrossRef]
9. Ratnakaram, Y.C.; Thirupathi Naidu, D.; Vijaya Kumar, A.; Gopal, N.O. Influence of mixed alkalies on absorption and emission properties of Sm^{3+} ions in borate glasses. *Phys. B* **2005**, *358*, 296–307. [CrossRef]
10. Annapurna, K.; Dwivedi, R.N.; Kumar, A.; Chaudhuri, A.K.; Buddhudu, S. Temperature dependent luminescence characteristic of Sm^{3+}-doped silicate glass. *Spectrochim. Acta A Mol. Biomol. Spectrosc.* **2000**, *56*, 103–109. [CrossRef]
11. Jayasimhadri, M.; Chon, E.-J.; Jang, K.-W.; Lee, H.S.; Kim, S. Spectroscopic properties and Judd-Ofelt analysis of Sm^{3+} doped lead-germanate-tellurite glasses. *J. Phys. D Appl. Phys.* **2008**, *41*, 175101. [CrossRef]
12. Sharma, Y.K.; Surana, S.S.L.; Singh, R.K. Spectroscopic investigation and luminescence spectra of soda lime silicate glasses. *J. Rare Earths* **2009**, *27*, 773. [CrossRef]
13. Basavapoornima, C.; Jayasankar, C.K. Spectroscopic and photoluminescence properties of Sm^{3+} ions in Pb-K-Al-Na phosphate glasses for efficient visible lasers. *J. Lumin.* **2014**, *153*, 233–241. [CrossRef]
14. Isard, J.O. The mixed alkali effect in glass. *J. Non-Cryst. Solids* **1969**, *1*, 235–261. [CrossRef]
15. Day, D.E. Mixed alkali glasses—Their properties and uses. *J Non-Cryst. Solids* **1976**, *21*, 343–372. [CrossRef]
16. Kracek, F.C. The ternary system K_2SiO_3-Na_2SiO_3-SiO_2. *J. Phys. Chem.* **1932**, *36*, 2529–2542. [CrossRef]
17. Selvi, S.; Marimuthu, K.; Muralidharan, G. Structural and luminescence behavior of Sm^{3+} ions doped lead boro-telluro-phosphate glasses. *J. Lumin.* **2015**, *159*, 207–218. [CrossRef]
18. Muralidharan, M.N.; Rasmitha, C.A.; Rateesh, R. Photoluminescence and FTIR studies of pure and rare earth doped silica xerogels and aerogels. *J. Porous Mater.* **2009**, *16*, 635–640. [CrossRef]
19. Sanders, D.M.; Person, W.B.; Hench, L.L. Quantitative-analysis of glass structure with the use of infrared reflection spectra. *Appl. Spectrosc.* **1974**, *28*, 247–255. [CrossRef]
20. Matson, D.W.; Sharma, S.K.; Philpotts, J.A. The structure of high-silica alkali-silicate glasses—A Raman spectroscopy investigation. *J. Non-Crystal. Solids* **1983**, *58*, 323–352. [CrossRef]
21. El-Okr, M.; Ibrahem, M.; Farouk, M. Structure and properties of rare-earth-doped glassy systems. *J. Non-Cryst. Solids* **2008**, *69*, 2564–2567. [CrossRef]
22. Srinvasa Rao, C.; Srikumar, T.; Rao, M.C. Physical and optical absorption studies on LIF/NaF/KF- P_2O_5-B_2O_3 glasses doped with Sm_2O_3. *IJCRGG* **2014**, *7*, 420–425.
23. Bhatia, B.; Meena, S.L.; Parihar, V.; Poonia, M. Optical basicity and polarizability of Nd^{3+}-doped bismuth borate glasses. *New J. Glass Ceram.* **2015**, *5*, 44–52. [CrossRef]
24. Judd, B.R. Optical absorption intensities of rare earth ions. *Phys. Rev.* **1962**, *127*, 750–761. [CrossRef]
25. Ofelt, G.S. Intensities of crystal spectra of rare earth ions. *J. Chem. Phys.* **1962**, *37*, 511–520. [CrossRef]

26. Jayasimhadri, M.; Moorthy, L.R.; Saleem, S.A.; Ravikumar, R.V.S.S. Spectroscopic characteristics of Sm^{3+}-doped alkali fluorophosphate glasses. *Spectrochim. Acta Part A* **2006**, *64*, 939–944. [CrossRef]
27. Jayasankar, C.K.; Rukmini, E. Optical properties of Sm^{3+} ions in zinc and alkali zinc borosulphate glasses. *Opt. Mater.* **1997**, *8*, 193–205. [CrossRef]
28. Chen, B.J.; Shen, L.F.; Pun, E.Y.B.; Lin, H. Sm^{3+}-doped germanate glass channel waveguide as light source for minimally invasive photodynamic therapy surgery. *Opt. Express* **2012**, *20*, 879–889. [CrossRef]
29. Jorgenson, C.K. *Orbitals Atoms and Molecules*; Academic Press: London, UK, 1962.
30. Sinha, S.P. *Complexes of the Rare Earth*; Pergamon Press: Oxford, UK, 1966.
31. Carnall, W.T.; Crosswhite, H.; Crosswhite, H.M. *Energy Level Structure and Transition Probabilities of the Trivalent Lanthanides in LaF3*; Report ANL–78–XX–95; Argonne National Laboratory: Lemont, IL, USA, 1978.
32. Sudhakar, K.S.V.; Srinivasa Reddy, M.; Srinivasa Rao, L.; Veeraiah, N. Influence of modifier oxide on spectroscopic and thermoluminescence characteristics of Sm^{3+}ion in antimony borate glass system. *J. Lumin.* **2008**, *128*, 1791–1798. [CrossRef]
33. Rajesh, D.; Balakrishna, A.; Ratnakaram, Y.C. Luminescence, structural and dielectric properties of Sm^{3+} impurities in strontium lithium bismuth borate glasses. *Opt. Mater.* **2012**, *35*, 108–116. [CrossRef]
34. Ratnakaram, Y.C.; Thirupathi Naidu, D.; Chakradhar, R.P.S. Spectral studies of Sm^{3+} and Dy^{3+} doped lithium cesium mixed alkali borate glasses. *J. Non-Cryst. Solids* **2006**, *353*, 3914–3922. [CrossRef]
35. Takebe, H.; Nageno, Y.; Morinaga, K. Compositional dependence of Judd-Ofelt parameters in silicate, borate, and phosphate glasses. *J. Am. Ceram. Sot.* **1995**, *78*, 1161–1168. [CrossRef]
36. Shanmuga Sundari, S.; Marimuthu, K.; Sivraman, M.; Surendra Babu, S. Composition dependent structural and optical properties of Sm^{3+}-doped sodium borate and sodium fluoroborate glasses. *J. Lumin.* **2010**, *130*, 1313–1319. [CrossRef]
37. Arul Rayappan, I.; Selvaraju, K.; Marimuthu, K. Structural and luminescence investigations on Sm^{3+} doped sodium fluoroborate glasses containing alkali/alkaline earth metal oxides. *Phys. B* **2011**, *406*, 548–555. [CrossRef]
38. Herrera, A.; Fernandes, R.G.; de Camargo, A.S.S.; Hernandes, A.C.; Buchner, S.; Jacinto, C.; Balzaretti, N.M. Visible–NIR emission and structural properties of Sm3+ doped heavy-metal oxide glass with composition B_2O_3–PbO–Bi_2O_3–GeO_2. *J. Lumin.* **2016**, *171*, 106–111. [CrossRef]
39. Montoya, I.M.; Balzaretti, N.M. High pressure effect on structural and spectroscopic properties of Sm^{3+} doped alkali silicate glasses. *High-Press. Res.* **2017**, *37*, 296–311. [CrossRef]

Disclaimer/Publisher's Note: The statements, opinions and data contained in all publications are solely those of the individual author(s) and contributor(s) and not of MDPI and/or the editor(s). MDPI and/or the editor(s) disclaim responsibility for any injury to people or property resulting from any ideas, methods, instructions or products referred to in the content.

Article

Bismuth-Germanate Glasses: Synthesis, Structure, Luminescence, and Crystallization

Ksenia Serkina [1], Irina Stepanova [1], Aleksandr Pynenkov [2], Maria Uslamina [2], Konstantin Nishchev [2], Kirill Boldyrev [3], Roman Avetisov [1] and Igor Avetissov [1,*]

[1] Department of Chemistry and Technology of Crystals, D. Mendeleev University of Chemical Technology of Russia (MUCTR), 125480 Moscow, Russia; serkina.k.v@muctr.ru (K.S.); stepanova.i.v@muctr.ru (I.S.); armoled@mail.ru (R.A.)
[2] Institute of High Technologies and New Materials, National Research Mordovia State University, 430005 Saransk, Russia; alekspyn@yandex.ru (A.P.); uslaminam@mail.ru (M.U.); nishchev@inbox.ru (K.N.)
[3] Laboratory of Fourier-Spectroscopy, Institute for Spectroscopy RAS, 108840 Troitsk, Russia; kn.boldyrev@gmail.com
* Correspondence: avetisov.i.k@muctr.ru

Abstract: Bismuth-germanate glasses, which are well known as a promising active medium for broadband near-infrared spectral range fiber lasers and as an initial matrix for nonlinear optical glass ceramics, have been synthesized in a 5–50 mol% Bi_2O_3 wide concentration range. Their structural and physical characteristics were studied by Raman and FT-IR spectroscopy, differential scanning calorimetry, X-ray diffraction, optical, and luminescence methods. It has been found that the main structural units of glasses are $[BiO_6]$ and $[GeO_4]$. The growth in bismuth oxide content resulted in an increase in density and refractive index. The spectral and luminescent properties of glasses strongly depended on the amount of bismuth active centers. The maximum intensity of IR luminescence has been achieved for the $5Bi_2O_3$-$95GeO_2$ sample. The heat treatment of glasses resulted in the formation of several crystalline phases, the structure and amount of which depended on the initial glass composition. The main phases were non-linear Bi_2GeO_5 and scintillating $Bi_4Ge_3O_{12}$. Comparing with the previous papers dealing with bismuth and germanium oxide-based glasses, we enlarge the range of Bi_2O_3 concentration up to 50 mol% and decrease the synthesis temperature from 1300 to 1100 °C.

Keywords: bismuth-germanate glass; bismuth active centers; IR-luminescence; crystallization

1. Introduction

The Bi_2O_3-GeO_2 system has a wide glass transition region up to 85.7 mol% Bi_2O_3 [1,2]. The basis of the structural network of bismuth-germanate glasses is $[GeO_4]^{4-}$-tetrahedra [3]. Bismuth oxide, as a modifier, creates additional bonds in glass, strengthening it. It is generally accepted that Bi^{3+} ions are predominantly in octahedral coordination in glass, but it can be varied from octahedral $[BiO_6]^{6-}$ to pyramidal $[BiO_3]^{3-}$ at Bi-concentration growth. Along with Bi^{3+}, bismuth can exist in other charge states in glasses [4].

Bismuth-containing glasses have a high refractive index and high density; they are transparent in the visible and IR spectral ranges [5,6]. Increased researchers' attention to these glasses arose after the discovery of a unique luminescence in the 1100–1500 nm range, the source of which is bismuth active centers (BACs) [7]. The structure of these centers has been subjected to changes over the past 20 years [8–10]. Up to date, the prevailing opinion is that BAC has a complex active structure, which is a combination of bismuth ions in low oxidation states and oxygen vacancies [11]. Understanding the nature of these centers would make it possible to optimize laser active media for the near-IR range.

In addition to active optics, bismuth-germanate glasses are used to produce glass-ceramic materials since the glass formation region of the Bi_2O_3-GeO_2 system includes

the compositions of different crystalline phases: "metastable" Bi_2GeO_5 with ferroelectric characteristics [12], and $Bi_4Ge_3O_{12}$ with scintillation properties [13].

All this makes bismuth-germanate glasses both unique and multipurpose materials. Thus, the goal of the present research was to investigate the properties of bismuth-germanate glasses synthesized in a wide range of bismuth and germanium oxide concentrations for further application in various fields of science and technology.

2. Materials and Methods

We synthesized $xBi_2O_3(100-x)GeO_2$ glasses in the 5–50 mol% Bi_2O_3 concentration range with a 5 mol% step. We used Bi_2O_3 99.999 wt% and GeO_2 99.995 wt% purchased from LANHIT LTD (Russia, Moscow). For a better presentation of the results, the samples were signed as x(100-x). For example, the $15Bi_2O_3$-$85GeO_2$ composition was signed as 15-85 (Sample ID). Glasses were synthesized in corundum crucibles at 1100 °C for 30 min by the standard melt-quenching technique with casting onto a metal substrate at room temperature. Thermal stresses were removed at 350 °C for 3 h, followed by cooling at a rate of ~50 °C/h. Polished parallel plates with 2 mm thickness were made from glasses for further studies.

The elemental analysis of the synthesized glasses was carried out using an X-ray spectral energy-dispersive microanalyzer (EDS Oxford Instruments X-MAX-50) on the base of a Tescan VEGA3-LMU scanning electron microscope (TESCAN ORSAY HOLDING, Brno, Czech Republic). Raman spectra were recorded on a QE65000 spectrophotometer (Ocean Optics, Largo, FL, USA) using a 785 nm excitation laser in the frequency shift range of 200–2000 cm^{-1} in backscattering geometry. IR transmission spectra were recorded on a Tensor 27 IR-Fourier spectrometer (Bruker, Ettlingen, Germany) in the 400–8000 cm^{-1} range.

The characteristic temperatures of the samples were determined by different scanning calorimetry methods using a DSC 404 F1 Pegasus instrument (Erich Netzsch GmbH & Co. Holding KG, Selb, Germany). The measurements were carried out for 100–120 mg samples placed in platinum crucibles at a 20 mL/min airflow and 10 °C/min heating rate. The density of the samples was determined by the hydrostatic method using a M-ER123 ACF JR-150.005 TFT balance (Mercury WP Tech Group Co., Ltd., Incheon, Republic of Korea) with an accuracy of 0.005 g/cm^3. The refractive index ($n_D > 1.78$) was determined using a MIN-8 optical microscope (LOMO JSC, Saint Peterburg, Russia) by measuring the shift of the refracted beam at different preset tilt angles of sample plates located on a special stage.

The absorption spectra were recorded on a UNICO 2800 (UV/VIS) spectrophotometer (United Products & Instruments, Suite E Dayton, NJ, USA) in the 190–1100 nm wavelength range with a 1 nm step. The luminescence spectra were recorded on an IFS 125HR FT-IR spectrometer (Bruker, Ettlingen, Germany) using an original self-made luminescent module. Luminescence was excited by diode lasers (CNI, Changchun, China) with wavelengths of 405, 425, 525, 650, and 805 nm; the power density on the sample was 100 mW/mm^2, and the spectral resolution was 12 cm^{-1}.

The original glasses were subjected to heat treatment for 2 h at various temperatures based on the DSC data. The structure of the crystalline phases was determined by X-ray diffraction using an Equinox-2000 diffractometer (Inel SAS, Artenay, France) with a linear CCD detector with a step of 0.0296 degrees in the range from 0 to 114 2θ and a 2400 s acquisition time using CuKα radiation (λ = 1.54056 Å). The phases were identified by a Match! Software package (2003–2015 CRYSTAL IMPACT, Bonn, Germany) as follows: α-GeO_2 (SG No 136; PDF #35-0729); α—GeO_2 (SG No 136; PDF #21-0902); β-GeO_2 (SG No 154; PDF #43-1016); $Bi_2Ge_3O_9$ (SG No 215; PDF #43-0216); $Bi_2Ge_3O_9$ (SG No 176; PDF #43-0216); $Bi_4Ge_3O_{12}$ (SG No 220; PDF #34-0416); Bi_2GeO_5 (SG No 36; PDF #36-0289); and $Bi_{12}GeO_{20}$ (SG No 197; PDF #77-0556).

3. Results

3.1. Glass Samples

The synthesized samples had a ruby-red color, which became more saturated with an increase in the bismuth oxide content (Figure 1). Samples containing >40 mol% Bi_2O_3 had color inhomogeneity, probably caused by heterogeneous component distribution in the glass.

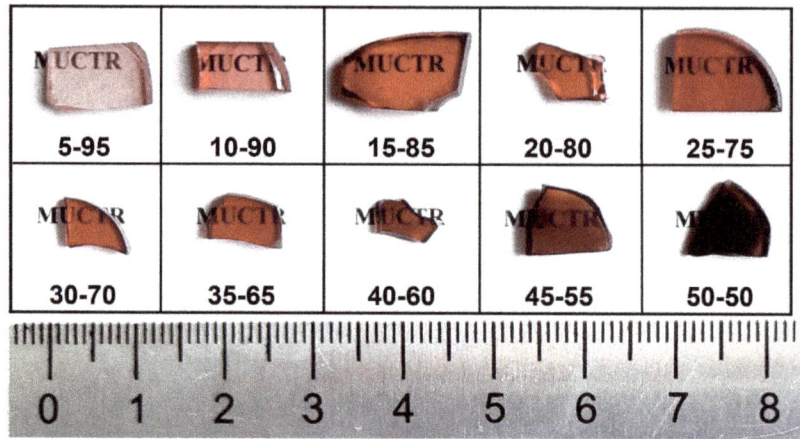

Figure 1. Photos of synthesized glasses. Here and after the numbers, refer to the sample ID.

The sample with the lowest content of bismuth oxide (5 mol%) had inclusions of bubbles due to the high melt viscosity at the synthesis temperature, and some of its properties were not studied. As a result, the 5-95 sample density was lower than the density of pure GeO_2. Glasses 50-50 were inclined to surface crystallization during melt casting, which contradicted the data of [2], in which $85.7Bi_2O_3$-$14.3GeO_2$ glasses were presented and similar synthesis conditions (temperature 1100–1200 °C, quenching on a metal substrate at room temperature) were reported for their production.

The results of the elemental analysis of the glasses showed that all samples contained an aluminum impurity (Table 1) due to the synthesis in corundum crucibles. A similar result was observed in [14]. With an increase in the bismuth oxide content, the amount of aluminum in the glass composition increased, which was explained by the chemical aggressiveness of the bismuth oxide melt towards the crucible material. Additionally, bismuth volatilized insignificantly during the synthesis, which was also described in the literature [15]. At the same time, the Bi/Ge ratio in our initial mixture and in the synthesized glass remained nearly unchanged.

Table 1. The composition of glasses according to the EDS data.

Sample ID	Bi Content (mol%)		Ge Content (mol%)		O Content (mol%)		Al Content (mol%)	
	Raw	EDS	Raw	EDS	Raw	EDS *	Raw	EDS
10-90	6.25	8.51 ± 0.29	28.13	25.89 ± 0.13	65.62	65.18	0.00	0.42 ± 0.02
15-85	9.09	10.91 ± 0.37	25.76	22.35 ± 0.11	65.15	63.25	0.00	3.49 ± 0.14
20-80	11.76	10.52 ± 0.36	23.53	21.77 ± 0.11	64.71	65.97	0.00	1.74 ± 0.07
25-75	14.29	13.54 ± 0.46	21.43	18.21 ± 0.09	64.28	63.62	0.00	4.63 ± 0.18
30-70	16.67	15.43 ± 0.52	19.45	16.36 ± 0.08	63.88	63.30	0.00	4.90 ± 0.19
35-65	18.92	17.01 ± 0.58	17.57	14.22 ± 0.07	63.51	62.84	0.00	5.93 ± 0.23
40-60	21.05	18.61 ± 0.63	15.79	12.78 ± 0.07	63.15	62.58	0.00	6.03 ± 0.24
45-55	23.08	20.93 ± 0.71	14.10	11.35 ± 0.06	62.82	62.22	0.00	5.50 ± 0.21
50-50	25.00	22.80 ± 0.78	12.50	9.76 ± 0.05	62.50	61.90	0.00	5.54 ± 0.22

* Oxygen content was calculated as 100-xBi-yGe-zAl.

3.2. Glass Structure Characterization

Structural units in the glass network were characterized using Raman and IR spectroscopy (Figures 2 and 3). In the low-frequency region (<700 cm^{-1}) for the Raman spectra (Figure 2), the bands in the region of 500 cm^{-1} characterized [GeO$_4$]-tetrahedra vibrations. Their intensity decreased with a reduction in germanium oxide concentration [2,16]. Additionally, in this region, there was a wide band at 600 cm^{-1}, related to vibrations of the Bi–O bond of [BiO$_6$]-octahedra [17]. It is interesting that for the 50-50 glass, the band at 395 cm^{-1}, associated with the bending of the O–Ge–O bridge bond [16], had the highest intensity in comparison with other glasses. It can be explained by the tendency of this glass to surface crystallize GeO$_2$ phases.

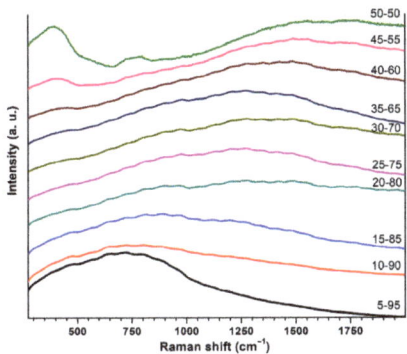

Figure 2. Raman spectra of synthesized glasses.

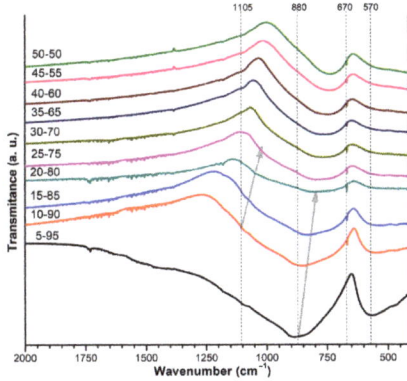

Figure 3. FT-IR spectra of synthesized glasses.

The bands in the high frequency region of the Raman spectra (>700 cm^{-1}) were assigned to [GeO$_4$]-tetrahedra vibrations with different numbers of non-bridging oxygen atoms, so-called Q$_n$-units, where n is the number of bridging oxygen atoms [18,19]. The growth of bismuth oxide content (Figure 2) resulted in the increasing intensity of the bands in the high-frequency region. This indicated an increase in the defectiveness of the glass structure.

The FT-IR spectra of the glasses (Figure 3) contained the main bands at 580, 670, 850, and 1105 cm^{-1}. The band at 580 cm^{-1} referred to asymmetric stretching of the Ge–O–Ge bridge bond vibrations [19] and was observed for all glasses; its intensity decreased with increasing Bi$_2$O$_3$ content. The band at 670 cm^{-1} was assigned to vibrations of Bi–O bonds in [BiO$_6$] structural units [20]. The band at 880 cm^{-1} was assigned to Ge–O–Ge stretching [21].

The band at 1105 cm^{-1} was assigned to vibrations of the Bi–O–Bi or Bi–O–Ge bond [20]. It should be noted that the bands at 880 and 1105 cm^{-1} shifted to the low-frequency region with an increase in the bismuth oxide content, which indicated a weakening of the Ge–O bonds due to the incorporation of bismuth ions into the glass network. The FT-IR transmission spectra (Figure 3) confirmed the assumptions about the glass structure and were in agreement with the Raman spectra presented above.

3.3. DSC Characterization and Physical Properties

The glass transition temperatures (T_g) and maximum crystallization temperatures (T_x) of all samples (Table 2) were determined from DSC curves (Figures S1–S10).

Table 2. Glass characteristic temperatures *.

Sample ID	T_g, °C	T_{x1}, °C	T_{x2}, °C	T_{x3}, °C
5-95	470	631	662	690
10-90	460	651	719	–
15-85	461	650	707	–
20-80	469	696	744	–
25-75	470	663	689	721
30-70	473	633	712	–
35-65	478	647	663	692
40-60	469	624	657	–
45-55	441	518	575	654
50-50	450	548	598	657

*—the determination error for all characteristic temperatures was ±1 °C.

The presence of several crystallization temperatures was associated with the formation of various crystalline phases. The difference in the number of crystallization temperatures (2 or 3) for different compositions can be associated both with a change in the type of crystallizing phases and with a rather high heating rate of the samples during the DSC processing. The formation of the metastable Bi_2GeO_5 phase could be observed in the 600–650 °C temperature range, according to [22]. The crystallization temperature in the region of 650–700 °C may correspond to the transition of the metastable Bi_2GeO_5 phase to the stable $Bi_4Ge_3O_{12}$ with the eulytite structure [23]. The shift of the crystallization temperatures of the same phase towards high values for glasses with a Bi_2O_3 content <30 mol% is explained by the lower tendency of these glasses to crystallize (Figure 4).

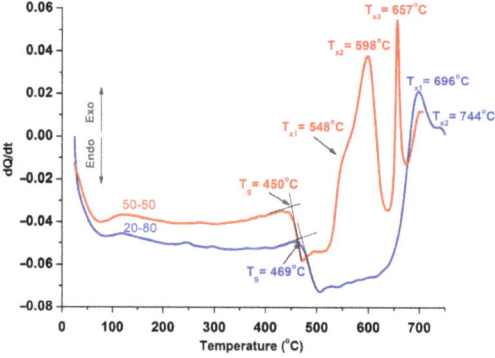

Figure 4. DSC curves of 20-80 and 50-50 synthesized glasses.

The density and refractive index of glasses (Table 3) expectedly increased with the growth in bismuth oxide content. The obtained results correlated with the data [5,6]. A slight decrease in the density and refractive index can be explained by the entry of aluminum oxide from the crucibles into the glasses.

Table 3. Density and refractive index values of glasses.

Sample	Density (g/cm³)	Refractive Index at 589 nm
0-100 *	4.25	
5-95	4.085 ± 0.005	1.68 ± 0.01
10-90	4.645 ± 0.005	1.76 ± 0.01
15-85	5.185 ± 0.005	1.80 ± 0.04
20-80	5.370 ± 0.005	1.86 ± 0.06
25-75	5.945 ± 0.005	2.06 ± 0.06
30-70	6.015 ± 0.005	2.08 ± 0.02
35-65	6.330 ± 0.005	2.10 ± 0.04
40-60	6.685 ± 0.005	2.12 ± 0.02
45-55	6.760 ± 0.005	2.14 ± 0.04
50-50	7.085 ± 0.005	2.14 ± 0.04
100-0 *	8.90	

* Values are presented for pure oxides.

3.4. Spectral-Luminescent Properties

The absorption spectra of glasses (Figure 5) exhibited a characteristic shoulder at 500 nm associated with BACs [7–9]. The absorption coefficient in this region increased with the growth of the bismuth oxide content.

Similarly, with an increase in the bismuth oxide content, the short-wavelength absorption edge shifted from 340 nm (Sample ID 5-95) to 425 nm (Sample ID 50-50). This shift was due to the fact that the optical band gap of bismuth (III) oxide is smaller than that of germanium oxide (5.63 eV) and ranges from 2.5 to 3.2 eV for various Bi_2O_3 polymorphs [14].

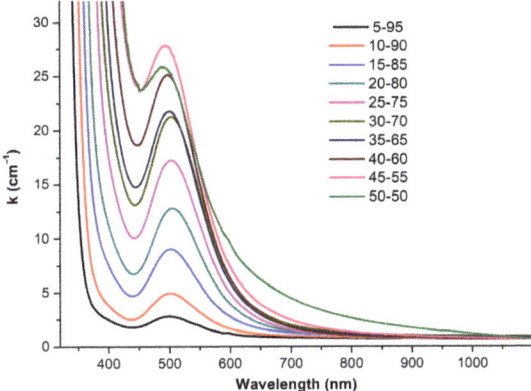

Figure 5. Optical absorption spectra of synthesized glasses.

To determine the width of the optical energy gap (E_g) of glasses, the Tauc method was used (Figure 6, Table 4).

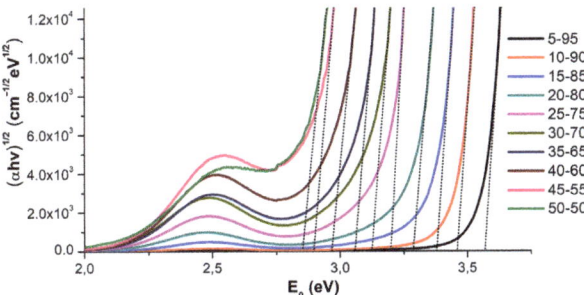

Figure 6. Tauc's plots.

Table 4. Energy band gap of bismuth-germanate glasses.

Sample ID	E$_g$ (eV) *
5-95	3.54
10-90	3.47
15-85	3.35
20-80	3.28
25-75	3.20
30-70	3.08
35-65	3.04
40-60	2.95
45-55	2.86
50-50	2.82

*—the determination error for E$_g$ was ±0.02 eV.

Under excitation of photoluminescence (PL) at wavelengths of 405, 425, 525, 650, and 805 nm for 5-95 samples (Figure 7), it was found that 450 nm was the optimal excitation for BACs (see Figure S19). We observed that a green laser (525 nm) action led to strong heating of the glasses, which significantly decreased the PL intensity.

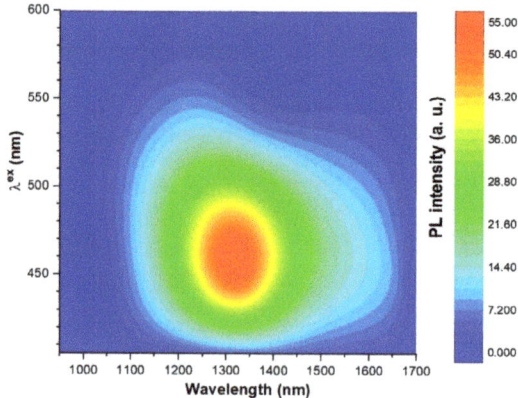

Figure 7. Luminescence spectrum of 5-95 glass at λ^{ex} = 400–600 nm.

The PL spectra of glasses at λ^{ex} = 450 nm (Figure 8) represented a wide band in the near IR region. As can be seen, the luminescence region corresponded to the data of [7–10], which additionally confirms the presence of BACs in glasses. For tested glasses, when the bismuth oxide content increased, the PL intensity became lower due to concentration quenching.

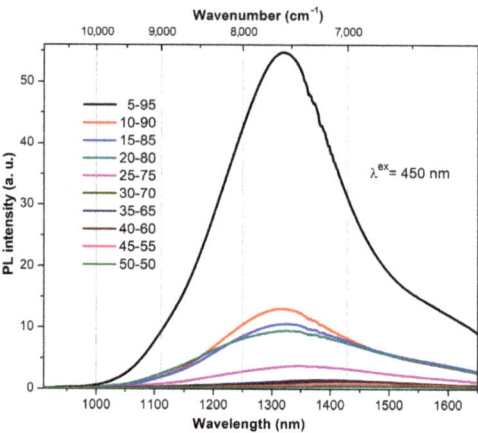

Figure 8. Photoluminescence spectra of glasses (λ^{ex} = 450 nm).

Sample 5-95 demonstrated the highest PL intensity (Figure 9). The observed broadband luminescence was attributed to low-valence forms of bismuth ($Bi^{n<2+}$) in the BACs [11,24,25].

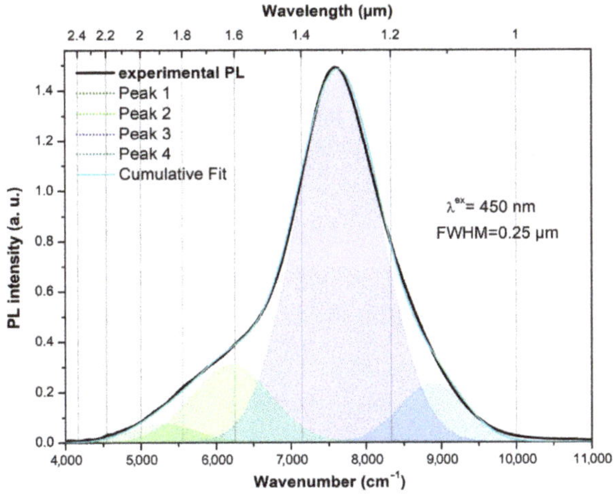

Figure 9. Photoluminescence spectrum of 5-95 glass (λ^{ex} = 450 nm).

4. Discussion

Analysis of the optical absorption and luminescence spectra of the synthesized glasses (λ^{ex} = 450 nm) showed the presence of BACs, the number of which increased with the bismuth oxide total concentration growth. The contour of the PL spectrum in the IR region was represented by a superposition of several bands, whose maxima, determined from the Gaussian components, were located at wavelengths ~1125, 1310, 1615, and 1885 nm (Figure 9). According to [26,27], the bands at 1125 and 1310 nm corresponded to the $^3P_1 \rightarrow {}^3P_0$ transitions for the Bi^+ ion and the $^2D_{3/2} \rightarrow {}^4S_{3/2}$ transitions for Bi^0, respectively.

At the same time, it was shown in [11] that BACs were not individual low-valence bismuth ions, but a complex system of cations and an oxygen vacancy. In this case, both bands at 1125 nm (Peak 4 in Figure 9) and 1310 nm (Peak 3 in Figure 9) belonged to oxygen-deficient centers =Bi···Ge≡. Thus, the difference in the band position was caused by the presence or absence of aluminum ions in the second coordination sphere of the BACs, respectively [11]. Previously, for the samples with a high content of Bi_2O_3 (>20 mol%), the luminescence was observed in the longer wavelength part of the spectrum (1800–3000 nm). It was supposed that this luminescence could be attributed to the formation of $Bi_5{}^{3+}$ cluster centers [28] or oxygen-deficient centers =Bi···Bi= [11]. We assume that in our glasses two types of luminescent BACs were formed: namely, =Bi···Ge≡ (~1125 and ~1310 nm) (Peaks 4 and 3 in Figure 9) and =Bi···Bi= (1615 and 1885 nm) (Peaks 2 and 1 in Figure 9) in a smaller amount. Bi_2O_3 content growth led to an increase in the amount of =Bi···Bi= type centers and to PL decreasing in the ~1300 nm region. This BACs transformation was in good agreement with the structural analysis data. The shift of the vibration bands towards low frequencies at the bismuth oxide content growth indicated an increase in the Ge–O and Bi–O bond lengths, which in turn resulted in the formation of =Bi···Bi= centers having shorter bond lengths than the =Bi···Ge≡ centers [11].

The glass transition temperatures of bismuth-containing glasses were lower compared to the temperature of pure GeO_2 glass (519 °C [29]), probably due to a decrease in melt viscosity upon the introduction of Bi_2O_3. The resulting range of T_g values (440–480 °C) was in good agreement with the data previously reported [30–33].

The DSC data showed the possibility of crystallization of several phases in glasses, and the set of crystalline phases varied for different glass compositions. The heat treatment of samples at 600 °C showed (Figure 10 and Figures S11–S17) that predominantly α-GeO_2

and β-GeO$_2$ phases crystallized, accompanied by a certain amount of the Bi$_4$Ge$_3$O$_{12}$ phase in samples containing up to 20 mol% Bi$_2$O$_3$. The crystallization peaks of all phases for these compositions were weakly separated (Figures S1–S3), which indicated the almost simultaneous beginning of their crystallization process in glass. The simultaneous existence of both modifications of crystalline GeO$_2$ correlated well with the metastable phase diagram [3], in which ~600 °C served as the transition temperature between α-GeO$_2$ and β-GeO$_2$ polymorphs. The Bi$_2$GeO$_5$ phase, noted in the same phase diagram, was unstable in this concentration range, as shown in [33], and appeared only in trace amounts in the 5-95 sample according to XRD patterns (see Figure 10).

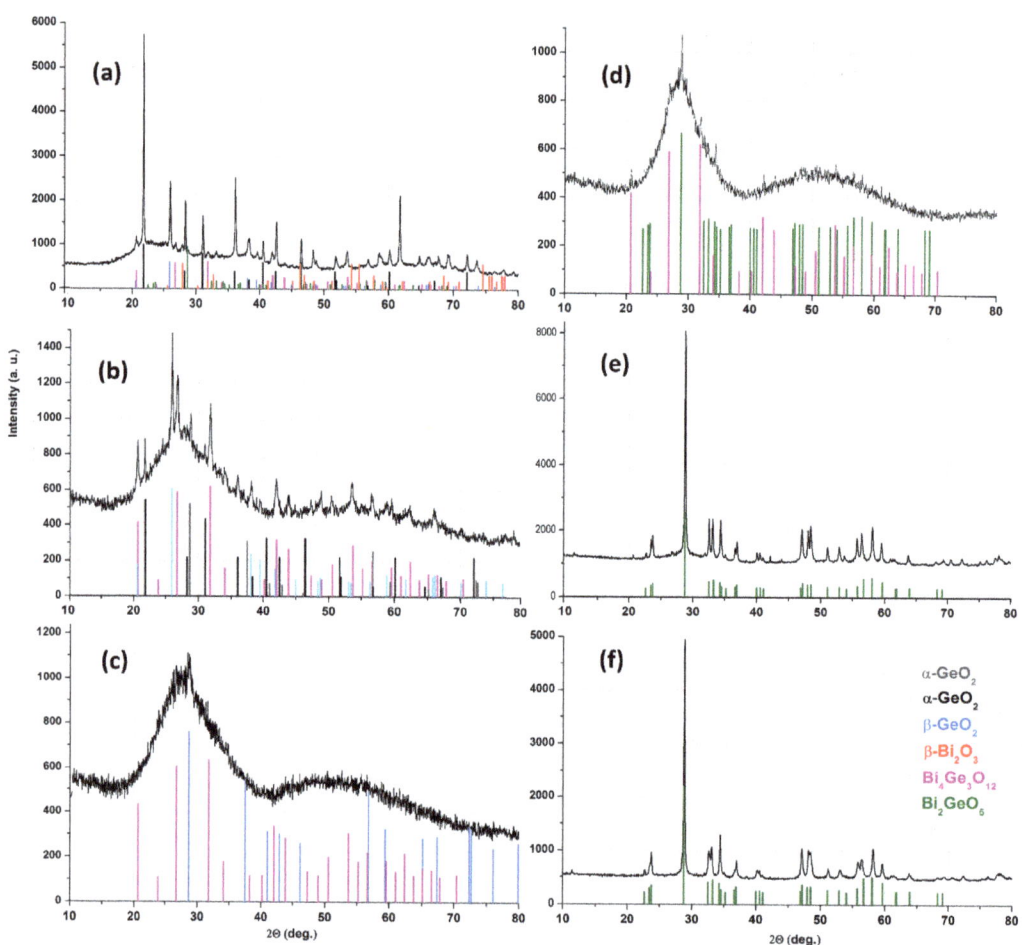

Figure 10. XRD patterns of crystallized glasses (a) 5-95, (b) 10-90, (c) 15-85, (d) 40-60, (e) 45-55, and (f) 50-50 heat-treated at 600 °C for 2 h (for details, see Figures S11–S13,S15–S17).

Crystallization in the 550–640 °C temperature range leads to the formation of the Bi$_4$Ge$_3$O$_{12}$ phase [30–32], alone or together with other phases for glass compositions containing 10–40 mol% Bi$_2$O$_3$. Therefore, the crystallization peaks belonging to the 650–663 °C range can be associated with the maximum crystallization temperature of the Bi$_4$Ge$_3$O$_{12}$ phase (Figure 11). The crystallization peaks in the 690–744 °C range can be associated with the crystallization temperature of the β-GeO$_2$ phase for samples with a high content of GeO$_2$ or other phases for samples with a high content of Bi$_2$O$_3$.

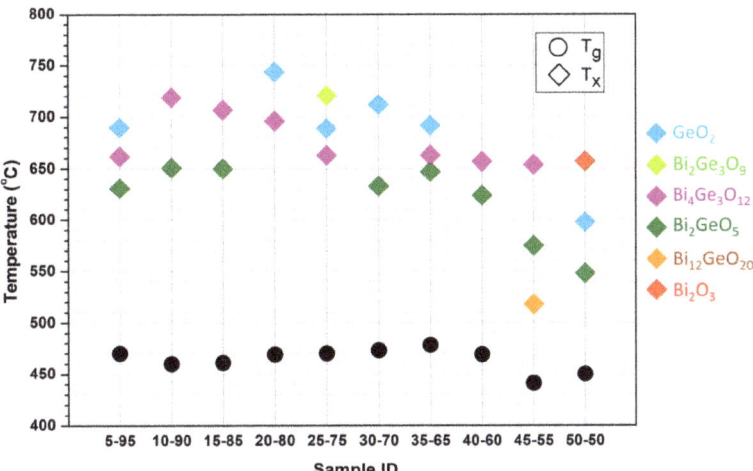

Figure 11. Glass transition and maximum crystallization temperatures of different phases formed from glasses with varied Bi_2O_3 composition.

It is known that the set of crystalline phases in crystallized glasses changes as the content of Bi_2O_3 increases (≥ 20 mol%). $Bi_4Ge_3O_{12}$ and Bi_2GeO_5 become the main phases [34,35]. Therefore, despite the XRD amorphous halo for our glasses containing 20–35 mol% Bi_2O_3 and heat-treated at 600 °C (Figure S14), it can be assumed that the crystallization proceeded similarly in our samples. Consequently, for the 624–647 °C range, the crystallization peaks belonged to the Bi_2GeO_5 phase. This assumption was also supported by the fact that the Bi_2GeO_5 phase disappearance was noted in [36] when the heat treatment temperature increased above 640 °C. Heat-treatment of 25-75 glass (Figure S18) at 690 °C led to the $Bi_4Ge_3O_{12}$ and $Bi_2Ge_3O_9$ phases' formation in agreement with [33,34]. An increase of the heat treatment temperature to 720 °C for the 25-75 glass led to an insignificant decrease in the amount of the $Bi_2Ge_3O_9$ phase, whose composition corresponded to that of glass; therefore, the crystallization maximum at 690 °C on the DSC curve corresponded to the $Bi_2Ge_3O_9$ phase formation. The formation of phases' mixtures during 25-75 glass crystallization corresponded to the cross sections of the Bi–Ge–O phase diagram [37], where in the region of 25 mol% Bi_2O_3 we observed $S_{Bi_4Ge_3O_{12}} - S_{Bi_2Ge_3O_9} - V$ bivariant equilibrium. The same phase equilibrium explains the absence of the β-GeO_2 phase in the crystallized glasses, which demonstrated a weak crystallization peak at 721 °C on the DSC curve.

The crystallization of 45-55 and 50-50 glasses should be discussed in detail. These glasses were inclined to crystallization in the glass casting process already. Therefore, there was a possibility of the spontaneous nuclei of crystalline phases' existence in glasses that were not determined by XRD. The crystallization maximum at temperatures of 575–598 °C belonged probably to the Bi_2GeO_5 phase since heat treatment at 600 °C led to the formation of this particular phase as the main one in 45-55 glass and the only one in 50-50 glass. To confirm these conclusions and to identify the crystallization peaks in the 518–548 °C range, which were not observed for the rest of the glass compositions, additional annealing of 45-55 and 50-50 glasses was carried out.

For the 45-55 sample (Figure 12), annealing at 520 °C led to the formation of the $Bi_{12}GeO_{20}$ phase together with the Bi_2GeO_5 and $Bi_4Ge_3O_{12}$ phases. The composition of the $Bi_{12}GeO_{20}$ phase corresponds to the molar composition of 85.7Bi_2O_3-14.3GeO_2, which is quite far from the original 45-55 glass composition. However, if we consider the Bi–Ge–O phase diagram cross sections [37] at temperatures close to the heat-treatment temperature (517 °C, 596 °C), it becomes clear that the 45-55 composition is in the range of monovariant equilibrium $S_{Bi_{12}GeO_{20}} - S_{Bi_4Ge_3O_{12}} - S_{Bi_2GeO_5} - V$. Annealing of 45-55 glass at 580 °C led to the crystallization of only two phases: Bi_2GeO_5 (basic) and $Bi_4Ge_3O_{12}$ (Figure 12). The area of $S_{Bi_{12}GeO_{20}} - S_{Bi_4Ge_3O_{12}} - S_{Bi_2GeO_5} - V$ monovariant equilibrium narrowed at temperature increases from 799 K to 850 K in the Bi–Ge–O diagram cross section [37]. As a result, the 45-55 glass composition moved to the region of $S_{Bi_4Ge_3O_{12}} - S_{Bi_2GeO_5} - V$ bivariant equilibrium. Thus, for 45-55 glass, the exothermic peak at 518 °C referred to the maximum crystallization temperature of the $Bi_{12}GeO_{20}$ phase, while the peak at 575 °C referred to the Bi_2GeO_5 phase.

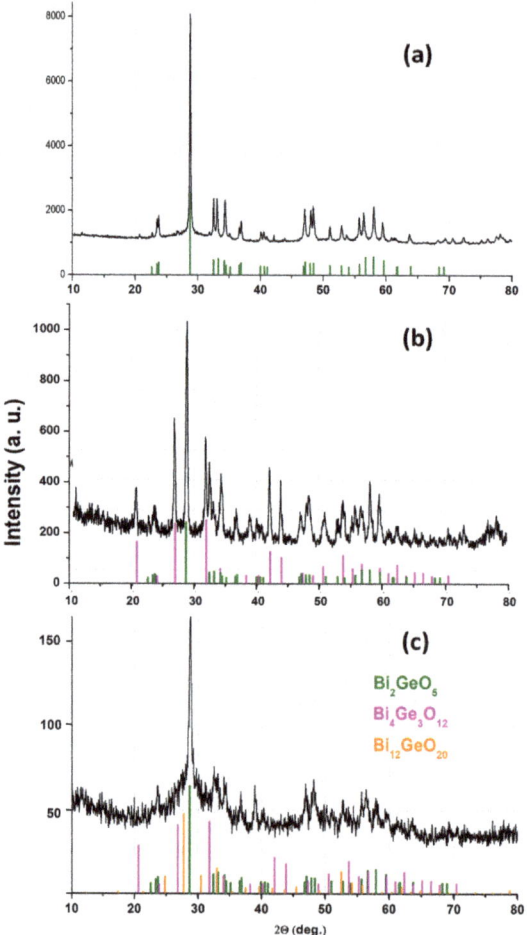

Figure 12. XRD patterns of crystallized 45-55 glasses heat-treated for 2 h at different temperatures: (**a**) 600 °C, (**b**) 580 °C, and (**c**) 520 °C.

Heat treatment of 50-50 glass at 550 °C led to the single Bi_2GeO_5 phase formation corresponding to the glass composition (Figure 13). The increase in heat-treatment temperature to 600 °C led to the appearance of the mixture of Bi_2GeO_5 and β-GeO_2 phases. The further temperature rise to 750 °C caused the formation of the mixture of Bi_2GeO_5, β-GeO_2, and β-Bi_2O_3 phases. Thus, the maximum crystallization temperatures of 548, 598, and 657 °C corresponded to the formation of Bi_2GeO_5, β-GeO_2, and β-Bi_2O_3 phases, respectively. The formation of the β-GeO_2 crystalline phase at high bismuth concentrations in the 50-50 sample could be caused by composition fluctuations in the initial glass.

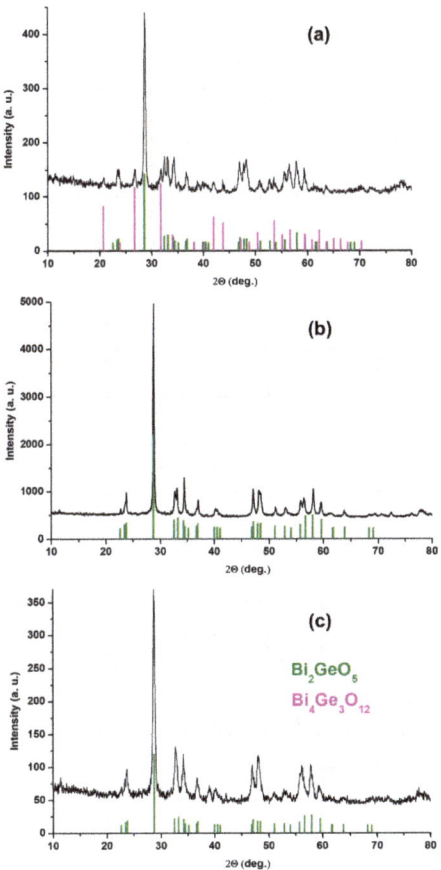

Figure 13. XRD patterns of crystallized 50-50 glasses heat-treated for 2 h at different temperatures: (a) 750 °C, (b) 600 °C, and (c) 550 °C.

Summarizing the crystallization data, we can say that the crystallization temperatures of bismuth-germanate phases correlate well with the amount of bismuth in their composition. The decrease in Bi_2O_3 content in the row of individual compounds $Bi_{12}GeO_{20} \rightarrow Bi_2GeO_5 \rightarrow Bi_4Ge_3O_{12} \rightarrow Bi_2Ge_3O_9$ (85.7–50–40–25 mol%) results in an increase in the maximum crystallization temperatures of the corresponding phases.

5. Conclusions

To fill the gaps in fundamental data for the first time, we investigated bismuth and germanium oxide-based glasses in a wide concentration range, with special emphasis on high Bi_2O_3 concentrations up to 50 mol%. We succeeded in decreasing the synthesis temperature from 1300 to 1100 °C. Glasses based on bismuth oxide and germanium oxide demonstrated a strong dependence of their structure and properties on the Bi_2O_3/GeO_2 ratio. An increase in the bismuth oxide concentration led to an increase in the number of non-bridging oxygen ions and a weakening of the Ge–O bonds. Such a rearrangement of the glass structure contributed to the destruction of =Bi···Ge≡ bismuth luminescent centers and the formation of =Bi···Bi= luminescent centers, which led to a weakening of the PL intensity in the region of ~1300 nm. The results of glass crystallization depended on the Bi_2O_3 oxide content: the higher the Bi_2O_3 concentration in a crystalline phase, the lower the temperature of its formation.

Supplementary Materials: The following supporting information can be downloaded at: https://www.mdpi.com/article/10.3390/ceramics6030097/s1, Figure S1: DSC curve of synthesized 5-95 glass; Figure S2: DSC curve of 10-90 glass; Figure S3: DSC curve of synthesized 15-85 glass; Figure S4: DSC curve of synthesized 20-80 glass; Figure S5: DSC curve of synthesized 25-75 glass; Figure S6: DSC curve of synthesized 30-70 glass; Figure S7: DSC curve of synthesized 35-65 glass; Figure S8: DSC curve of synthesized 40-60 glass; Figure S9: DSC curve of synthesized 45-55 glass; Figure S10: DSC curve of synthesized 50-50 glass; Figure S11: XRD patterns of 5-95 glass heat-treated at 600 °C; Figure S12: XRD patterns of 10-90 glass heat-treated at 600 °C; Figure S13: XRD patterns of 15-85 glass heat-treated at 600 °C; Figure S14: XRD patterns of 20-80, 25-75, 30-70, and 35-65 glasses heat-treated at 600 °C; Figure S15: XRD patterns of 40-60 glass heat-treated at 600 °C; Figure S16: XRD patterns of 45-55 glass heat-treated at 600 °C; Figure S17: XRD patterns of 50-50 glass heat-treated at 600 °C; Figure S18: XRD patterns of 25-75 glass heat-treated at 600, 690, and 720 °C; Figure S19: Excitation and emission spectra of 5-95 glass (λ^{ex} = 450 nm).

Author Contributions: Conceptualization, I.S. and I.A.; methodology, K.S.; software, K.N.; validation, K.S., I.S. and I.A.; formal analysis, M.U.; investigation, K.S., I.S., A.P., M.U., K.B. and R.A.; resources, K.N., K.B. and I.A.; data curation, K.S. and I.S.; writing—original draft preparation, K.S. and I.S.; writing—review and editing, I.S., R.A. and I.A.; visualization, K.S.; supervision, I.A.; project administration, I.S.; funding acquisition, R.A. All authors have read and agreed to the published version of the manuscript.

Funding: The research was financially supported by the Ministry of Science and Higher Education of Russia through the project FSSM-2020-0005.

Institutional Review Board Statement: Not applicable.

Informed Consent Statement: Not applicable.

Data Availability Statement: Not applicable.

Conflicts of Interest: The authors declare no conflict of interest.

References

1. Fedelesh, V.I.; Kutsenko, Y.P.; Turyanitsa, I.D.; Chepur, D.V. Elastooptic characteristics of glasses of the Bi_2O_3-GeO_2 system. *Fiz. Khim. Stekla* **1983**, *9*, 247–248.
2. Beneventi, P.; Bersani, D.; Lottici, P.P.; Kovács, L.; Cordioli, F.; Montenero, A.; Gnappi, G. Raman study of Bi_2O_3-GeO_2-SiO_2 glasses. *J. Non-Cryst. Solids* **1995**, *192–193*, 258–262. [CrossRef]
3. Zhereb, V.P.; Skorikov, V.M. Metastable states in bismuth-containing oxide systems. *Inorg. Mater.* **2003**, *39*, S121–S145. [CrossRef]
4. Maeder, T. Review of Bi_2O_3-based glasses for electronics and related applications. *Int. Mater. Rev.* **2012**, *58*, 3–40. [CrossRef]
5. Kusz, B.; Trzebiatowski, K. Bismuth germanate and bismuth silicate glasses cryogenic detectors. *J. Non-Cryst. Solids* **2003**, *319*, 257–262. [CrossRef]
6. Riebling, E.F. Depolymerization of GeO_2 and GeO_2·Sb_2O_3 glasses by Bi_2O_3. *J. Mater. Sci.* **1974**, *9*, 753–760. [CrossRef]
7. Fujimoto, Y.; Nakatsuka, M. Infrared luminescence from bismuth-doped silica glass. *J. Jpn. Appl. Phys.* **2001**, *40*, L279–L281. [CrossRef]
8. Meng, X.; Qiu, J.; Peng, M.; Chen, D.; Zhao, Q.; Jiang, X.; Zhu, C. Near infrared broadband emission of bismuth-doped aluminophosphate glass. *Opt. Express* **2005**, *13*, 1635–1642. [CrossRef]
9. Denker, B.; Galagan, B.; Osiko, V.; Sverchkov, S.; Dianov, E. Luminescent properties of Bi-doped boro-alumino-phosphate glasses. *J. Appl. Phys.* **2007**, *87*, 135–137. [CrossRef]
10. Hughes, M.; Akada, T.; Suzuki, T.; Ohishi, Y.; Hewak, D.W. Ultrabroad emission from a bismuth doped chalcogenide glass. *Opt. Express* **2009**, *17*, 19345–19355. [CrossRef]
11. Sokolov, V.O.; Plotnichenko, V.G.; Dianov, E.M. Origin of near-IR luminescence in Bi_2O_3-GeO_2 and Bi_2O_3-SiO_2 glasses: First-principle study. *Opt. Mater. Exp.* **2015**, *5*, 163–168. [CrossRef]
12. Pengpat, K.; Holland, D. Glass-ceramics containing ferroelectric bismuth germanate (Bi_2GeO_5). *J. Eur. Ceram. Soc.* **2003**, *23*, 1599–1607. [CrossRef]
13. Macedo, Z.S.; Silva, R.S.; Valerio, M.; Martinez, A.; Hernandes, A. Laser-sintered bismuth germanate ceramics as scintillator devices. *J. Am. Ceram. Soc.* **2004**, *87*, 1076–1081. [CrossRef]
14. Garcia dos Santos, M.; Moreira, R.C.M.; Gouveia de Souza, A.; Lebullenger, R.; Hernandes, A.C.; Leite, E.R.; Paskocimas, C.A.; Longo, E. Ceramic crucibles: A new alternative for melting of PbO-$BiO_{1.5}$-$GaO_{1.5}$ glasses. *J. Non-Cryst. Solids* **2003**, *319*, 304–310. [CrossRef]
15. Zhao, Y.; Wondraczek, L.; Mermet, A.; Peng, M.; Zhang, Q.; Qiu, J. Homogeneity of bismuth-distribution in bismuth-doped alkali germanate laser glasses towards superbroad fiber amplifiers. *Opt. Express* **2015**, *23*, 12423–12433. [CrossRef]
16. Henderson, G.S.; Wang, H.M. Germanium coordination and the germanate anomaly. *Opt. Mater.* **2014**, *5*, 163–168. [CrossRef]

17. Zhang, X.; Yin, S.; Wan, S.; You, J.; Chen, H.; Zhao, S.; Zhang, Q. Raman spectrum analysis on the solid–liquid boundary layer of BGO crystal growth. *Chin. Phys. Lett.* **2007**, *24*, 1898–1900.
18. Di Martino, D.; Santos, L.F.; Marques, A.; Almeida, R. Vibrational spectra and structure of alkali germanate. *J. Non-Cryst. Solids* **2001**, *293–295*, 394–401. [CrossRef]
19. Koroleva, O.N.; Shtenberg, M.V.; Ivanova, T.N. The structure of potassium germanate glasses as revealed by Raman and IR spectroscopy. *J. Non-Cryst. Solids* **2019**, *510*, 143–150. [CrossRef]
20. Pascuta, P.; Pop, L.; Rada, S.; Bosca, M.; Culea, E. The local structure of bismuth borate glasses doped with europium ions evidenced by FT-IR spectroscopy. *J. Mater. Sci.* **2008**, *19*, 424–428. [CrossRef]
21. Laudisio, G.; Catauro, M. The non-isothermal devitrification of $Li_2O \cdot TiO_2 \cdot 6GeO_2$ glass. *Thermochim. Acta* **1998**, *320*, 155–159. [CrossRef]
22. Cho, J.H.; Kim, S.J.; Yang, Y.S. Structural change in $Bi_4(Si_xGe_{1-x})_3O_{12}$ glasses during crystallization. *Solid State Commun.* **2001**, *119*, 465–470. [CrossRef]
23. Yu, P.; Su, L.; Cheng, J.; Zhang, X.; Xu, J. Study on spectroscopic properties and effects of tungsten ions in $2Bi_2O_3$-$3GeO_2/SiO_2$ glasses. *Appl. Radiat. Isot.* **2017**, *122*, 106–110. [CrossRef]
24. Veber, A.A.; Usovich, O.V.; Trusov, L.A.; Kazin, P.E.; Tsvetkov, V.B. Luminescence centers in silicate and germanate glasses activated by bismuth. *Bull. Lebedev Phys. Inst.* **2012**, *39*, 305–310. [CrossRef]
25. Dianov, E.M. On the nature of near-IR emitting Bi centres in glass. *Quantum Electron.* **2010**, *40*, 283–285. [CrossRef]
26. Wang, R.; Liu, J.; Zhang, Z. Luminescence and energy transfer progress in Bi-Yb co-doped germanate glass. *J. Alloys Compd.* **2016**, *688*, 332–336. [CrossRef]
27. Zhang, N.; Qiu, J.; Dong, G.; Yang, Z.; Zhang, Q.; Peng, M. Broadband tunable near-infrared emission of Bi-doped composite germanosilicate glasses. *J. Mater. Chem.* **2012**, *22*, 3154. [CrossRef]
28. Jiang, X.; Su, L.; Guo, X.; Tang, H.; Fan, X.; Zhan, Y.; Wang, Q.; Zheng, L.; Li, H.; Xu, J. Near-infrared to mid-infrared photoluminescence of Bi_2O_3-GeO_2 binary glasses. *Opt. Lett.* **2012**, *37*, 4260. [CrossRef]
29. Płonska, M.; Plewa, J. Crystallization of GeO_2-Al_2O_3-Bi_2O_3 glass. *Crystals* **2020**, *10*, 522. [CrossRef]
30. Rojas, S.S.; De Souza, J.E.; Andreeta, M.R.B.; Hernandes, A.C. Influence of ceria addition on thermal properties and local structure of bismuth germanate glasses. *J. Non-Cryst. Solids* **2010**, *356*, 2942–2946. [CrossRef]
31. Aldica, G.; Polosan, S. Investigations of the non-isothermal crystallization of $Bi_4Ge_3O_{12}$ (2:3) glasses. *J. Non-Cryst. Solids* **2012**, *358*, 1221–1227. [CrossRef]
32. Shi, Z.; Lv, S.; Tang, G.; Tang, J.; Jiang, L.; Qian, Q.; Zhou, S.; Yang, Z. Multiphase transition toward colorless bismuth-germanate scintillating glass and fiber for radiation detection. *ACS Appl. Mater. Interfaces* **2020**, *12*, 17752–17759. [CrossRef]
33. Bermeshev, T.V.; Zhereb, V.P.; Bundin, M.P.; Yasinsky, A.S.; Yushkova, O.V.; Voroshilov, D.S.; Zaloga, A.N.; Kovaleva, A.A.; Yakiv'yuk, O.V.; Samoilo, A.S.; et al. Synthesis of $Bi_2Ge_3O_9$. *Inorg. Mater.* **2022**, *58*, 1274–1283. [CrossRef]
34. Jiang, X.; Su, L.; Yu, P.; Guo, X.; Tang, H.; Xu, X.; Zheng, L.; Li, H.; Xu, J. Broadband photoluminescence of Bi_2O_3-GeO_2 binary systems: Glass, glass-ceramics and crystals. *Laser Phys.* **2013**, *23*, 105812. [CrossRef]
35. Gökçe, M.; Koçyiğit, D. Structural and optical properties of Gd^{3+} doped Bi_2O_3-GeO_2 glasses and glass-ceramics. *Mater. Res. Exp.* **2018**, *6*, 025203. [CrossRef]
36. Dimesso, L.; Gnappi, G.; Montenero, A.; Fabeni, P.; Pazzi, G.P. The crystallization behaviour of bismuth germanate glasses. *J. Mater. Sci.* **1991**, *26*, 4215–4219. [CrossRef]
37. Stepanova, I.V.; Petrova, O.B.; Korolev, G.M.; Guslistov, M.I.; Zykova, M.P.; Avetisov, R.I.; Avetissov, I.C. Synthesis of the Bi_2GeO_5 ferroelectric crystalline phase from a nonstoichiometric batch. *Phys. Status Solidi A* **2022**, *219*, 2100666. [CrossRef]

Disclaimer/Publisher's Note: The statements, opinions and data contained in all publications are solely those of the individual author(s) and contributor(s) and not of MDPI and/or the editor(s). MDPI and/or the editor(s) disclaim responsibility for any injury to people or property resulting from any ideas, methods, instructions or products referred to in the content.

Article

Formation and Photophysical Properties of Silver Clusters in Bulk of Photo-Thermo-Refractive Glass

Leonid Yu. Mironov [1,*], Dmitriy V. Marasanov [1], Mariia D. Sannikova [1], Ksenia S. Zyryanova [1], Artem A. Slobozhaninov [1] and Ilya E. Kolesnikov [2]

[1] Research and Educational Center for Photonics and Optoinformatics, ITMO University, Kronverkskiy Pr. 49, Saint Petersburg 197101, Russia
[2] Center for Optical and Laser materials research, St. Petersburg University, 7/9 Universitetskaya Nab, Saint Petersburg 199034, Russia
* Correspondence: leonid_mironov@itmo.ru

Abstract: The bright luminescence of silver clusters in glass have potential applications in solid-state lighting, optical memory, and spectral converters. In this work, luminescent silver clusters were formed in the bulk of photo-thermo-refractive glass ($15Na_2O$-$5ZnO$-$2.9Al_2O_3$-$70.3SiO_2$-$6.5F$, mol.%) doped with different Ag_2O concentrations from 0.01 to 0.05 mol.%. The spontaneous formation of plasmonic nanoparticles during glass synthesis was observed at 0.05 mol.% of Ag_2O in the glass composition, limiting the silver concentration range for cluster formation. The luminescence of silver clusters was characterized by steady-state and time-resolved spectroscopy techniques. The rate constants of fluorescence, phosphorescence, intersystem crossing, and nonradiative deactivation were estimated on the basis of an experimental study. A comparison of the results obtained for the photophysical properties of luminescent silver clusters formed in the ion-exchanged layers of photo-thermo-refractive glass is provided.

Keywords: glass; luminescence; silver clusters

1. Introduction

Functional optical materials based on silver clusters have found different applications in white light generation [1], optical data storage [2–4], sensing [5,6], spectral conversion [7], waveguides [8–10], and radiation measurements [11,12]. Currently, the development of new materials for effective solid-state lighting is attracting a lot of attention because of its potential to reduce electricity consumption. White light generation with silver clusters was achieved through the combination of a blue light-emitting diode (LED) with a layer of organic ligand-stabilized Ag_6 clusters with a quantum yield of 95% at room temperature [13]. Although the emission of Ag_6 clusters was shown to be thermally stable, the combined emission of blue LED and Ag_6 showed a significant drop near 500 nm, which is similar to the widely applied cerium-doped yttrium aluminum garnet (YAG:Ce) luminophore [14].

Luminescent silver clusters in inorganic hosts, such as zeolites and glasses, have great potential as light-emissive materials for white LEDs. Silver cluster-doped materials possess broad emission spectra, covering the whole visible spectrum under long-wave ultraviolet (UV) excitation around 365 nm, which is available from commercial LEDs. Unlike traditional luminophores, such as rare-earth-doped or transition metal-doped crystal powders, silver clusters have no distinguished emission bands, providing natural lighting with a high color-rendering index. Luminescent materials based on silver clusters in zeolites were shown to have intense green-yellow luminescence with a quantum yield of 83% in Linde Type A (LTA) zeolites [15–17] and up to 100% in faujasite (FAU) zeolites [18]. Although zeolite-based materials are established as promising materials for white light emission, their long-term stability may be affected by the sorption properties of zeolites.

It was shown that the external quantum efficiency of silver clusters in LTA, FAUX, and FAUY zeolites fell 1.1–10 times after one month in a high-humidity environment [19]. Unlike zeolites, inorganic glass provides excellent chemical stability for silver clusters dispersed in the bulk of glass. The use of oxyfluoride glass doped with silver was proposed to generate white light with different color temperatures under near-UV irradiation [20]. Recently, a prototype of a white LED was realized by a combination of a UV LED with peak emission at 365 nm and silver clusters dispersed in borosilicate glass. The prototype was characterized by color coordinates of (0.32, 0.37) and a color rendering index of 89.7, indicating that silver clusters dispersed in inorganic glass have the potential for white light generation [21].

Silicate glasses have been used to host luminescent silver clusters because of their chemical and mechanical stability. Also, it was shown that in silica-based glass, silver clusters can have a quantum yield of luminescence up to 63% [22]. There are three main methods to introduce silver ions into silicate glass: the addition of silver compounds to a glass batch, ion exchange, and ion implantation. The most straightforward approach is the addition of silver compounds to the glass batch, leading to the formation of silver clusters in the bulk of the material. Since the solubility of silver ions in silica glasses is low, the main limitation of this approach is the spontaneous aggregation of luminescent silver clusters into larger non-luminescent silver species, such as plasmonic nanoparticles [23]. The ion exchange method modifies the surface layer of the glass after synthesis; it is necessary to introduce alkali metal ions into the glass composition to exchange them with silver ions by immersing them in a silver-containing salt melt. It is possible to introduce >10 mol.% of Ag_2O into silicate glass by tuning the glass composition, the time and temperature of ion exchange, and the salt melt composition [24]. Additionally, ion exchange is an established method of optical waveguide production [25]. The thickness of an ion exchange layer depends on the exact parameters of the process and reaches several tens of micrometers. Similar to ion exchange, ion implantation also introduces silver ions only into the surface layer of the glass, but the thickness of the modified layer is usually less than two micrometers [26,27]. Also, ion implantation requires more complicated equipment in comparison with other methods.

In this work, we synthesized luminescent silver clusters in the bulk of photo-thermo-refractive glass and extensively studied their luminescent properties using steady-state and time-resolved spectroscopy. Earlier, the formation of luminescent silver clusters was shown in the bulk of glass with a similar composition, but the optical characterization lacked the study of luminescence quantum yields as well as fluorescence and phosphorescence decay kinetics [23]. Furthermore, the comparison of photophysical properties with the properties of silver clusters synthesized using the ion exchange technique in similar glass was performed. It was shown that although the concentration of Ag^+ was 200–500 times larger for the ion-exchanged glasses, the photophysical properties of the silver clusters were close. This result supports the suggestion that only small silver clusters, consisting of only 2–4 atoms, are luminescent.

2. Materials and Methods

Photo-thermo-refractive glass with the composition $15Na_2O$-$5ZnO$-$2.9Al_2O_3$-$70.3SiO_2$-$6.5F$ (mol.%) was used to form luminescent clusters in the bulk of the glass. The glass matrix was additionally doped with 0.02 mol.% of Sb_2O_3 and 0.007 mol.% of CeO_2. The concentrations of Ag_2O were 0.01, 0.025, and 0.05 mol.%, corresponding to the samples Ag.01, Ag.025, and Ag.05. The glass was synthesized in a quartz ceramic (stekrit) crucible at 1440 °C in an air atmosphere, and a platinum stirrer was used to homogenize the glass melt. The glass transition temperature (T_g) of the samples was measured with an STA 449 F1 Jupiter (Netzsch) differential scanning calorimeter at a heating rate of 10 K/min, and T_g was found to be 486 °C. Glass samples were cut in plates with a thickness of ~1 mm and further polished for spectroscopy analysis. To initiate cluster formation, the glass samples were irradiated with a mercury lamp for 10 min on each side and subsequently heat-treated. Irradiation was necessary to ionize Ce^{3+} into Ce^{4+}, and the released electrons were trapped by Sb^{5+} ions forming $(Sb^{5+})^-$ species. The time of irradiation was chosen to ensure the saturation of Ce^{3+} ionization. Further heat treatment provided energy to release electrons from $(Sb^{5+})^-$ and reduce Ag^+ ions to Ag^0, followed by cluster formation. The mechanism of photo-thermo-induced crystallization in the PTR glass has been studied in detail previously [28,29]. The heat treatment of the irradiated glass samples was carried out at 350, 400, and 450 °C for three hours.

Absorption spectra were measured using a PerkinElmer Lambda 650 spectrophotometer. Steady-state emission spectra and absolute quantum yields of luminescence were acquired using a Hamamatsu C9920 setup equipped with a 150 W CW xenon lamp, an 8.3 cm integrating sphere, and a PMA-12 CCD spectrometer. Time-resolved emission spectra were measured using a PerkinElmer LS50B fluorometer equipped with a pulsed xenon lamp with a pulse width at half maximum < 10 µs. The fluorometer was used in the phosphorescence measurement mode, which made it possible to record sample emission with a fixed delay time after the excitation pulse. Total emission spectra representing combined fluorescence and phosphorescence of silver clusters were measured without delay after the excitation pulse. Further, the phosphorescence of silver clusters was detected with a delay of 40 µs after the excitation pulse to ensure full decay of the excitation pulse; the value of the delay time was established according to the scattering reference. The fluorescence spectra were reconstructed by subtracting the phosphorescence spectra from the total emission spectra. Steady-state and time-resolved spectra were measured under 340 nm excitation; the second order diffraction peak from excitation pulse was blocked with a cut-off filter. All spectra were corrected for the sensitivity of the detector. Fluorescence decay curves were obtained with a HORIBA Fluorolog-3 fluorometer using the time-correlated single photon counting (TCSPC) technique, and fluorescence was excited by an LED at 340 nm with a 1.2 ns pulse. The experimental decay curves were fitted using bi-exponential or tri-exponential functions, depending on the goodness of the fit. Average fluorescence lifetimes were obtained using the following equation in the case of the bi-exponential fitting:

$$\tau_{avg} = \frac{\alpha_1 \tau_1^2 + \alpha_2 \tau_2^2}{\alpha_1 \tau_1 + \alpha_2 \tau_2}, \tag{1}$$

where α_i is the amplitude of the i-th component, and τ_i is the lifetime of the i-th component. Average fluorescence lifetimes in the case of triexponential fitting were calculated using the following equation:

$$\tau_{avg} = \frac{\alpha_1 \tau_1^2 + \alpha_2 \tau_2^2 + \alpha_3 \tau_3^2}{\alpha_1 \tau_1 + \alpha_2 \tau_2 + \alpha_3 \tau_3}, \tag{2}$$

where α_i is the amplitude of the i-th component, and τ_i is the lifetime of the ith component. Phosphorescence decay curves were obtained using phosphorescence spectra recorded with different delays after the excitation pulse and a fixed gate time.

3. Results and Discussion

Silica-based glasses have a low solubility of silver ions, which limits the concentration of Ag_2O introduced through the glass batch. Absorption spectra of the initial glass samples are shown in Figure 1a, demonstrating that the Ag.01 and Ag.025 samples were transparent in the visible range. A characteristic absorption band of Ce^{3+} ions was observed at 310 nm for all samples with the contribution of Ce^{4+} ions in the shorter wavelength region [30]. Additionally, the Ag.05 sample had an intense absorption peak at 410 nm corresponding to the plasmon absorption of silver nanoparticles. Therefore, spontaneous formation of silver nanoparticles during the synthesis of glass with the Ag_2O concentration of 0.05 mol.% was observed for the studied glass system. UV irradiation of the glass samples with a mercury lamp induced the appearance of additional unstructured absorption caused by the transition of cerium ions to the Ce^{4+} state (Figure 1b–d) and a decrease in the intensity of Ce^{3+} absorption bands as a result of photoionization. The increase in absorption originated from a tail of the strong charge transfer transition of Ce^{4+} in the near-UV region, since Ce^{4+} has a $4f^0$ configuration and, therefore, cannot exhibit f-f or f-d transitions [31]. Further heat treatment had a different effect on the samples with different Ag_2O concentrations. The Ag.01 sample demonstrated the formation of typical unstructured absorption of silver clusters between 320 and 400 nm after heat treatment at 350 °C (Figure 1b). An increase in the heat treatment temperature to 400 and 450 °C led to the appearance of distinguishable absorption peaks of silver nanoparticles. It is interesting that the Ag.025 sample with a higher silver concentration demonstrated the appearance of silver nanoparticles only after heat treatment at 450 °C (Figure 1c). The Ag.05 sample retained the shape and position of the plasmon peak after UV irradiation and heat treatment (Figure 1d). The formation of unstructured absorption overlapping with the plasmon peak is attributed to the appearance of luminescent silver clusters, since it was shown further that Ag.05 glass samples possess efficient emission after heat treatment.

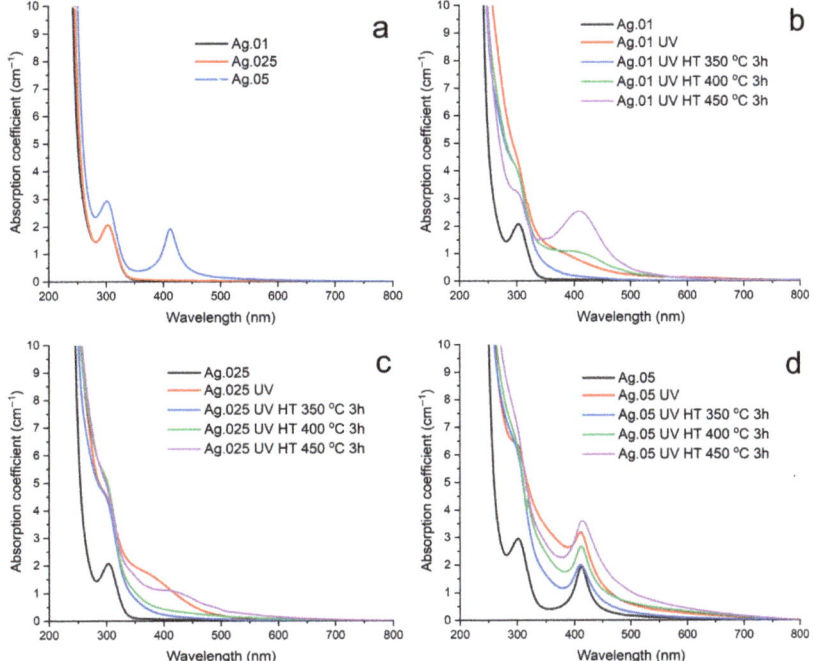

Figure 1. Comparison of initial glass absorption spectra (**a**). Influence of UV irradiation and heat treatment on the absorption of Ag.01 (**b**), Ag.025 (**c**), and Ag.05 (**d**) glass samples.

The calculation of the effective optical size of silver nanoparticles was carried out according to the Mie theory [32,33] using the following equation:

$$d = (2\nu F)/\Delta w, \qquad (3)$$

where d is the average silver nanoparticle diameter, νF is the Fermi velocity (1.39×10^8 cm/s for silver [34]), and Δw is the full width at half maximum (FWHM) in angular frequency units. The results of the nanoparticle size calculation for the samples with distinguishable plasmon peaks are shown in Table 1. It can be seen that the increase in the concentration of Ag_2O in glass composition led to the formation of larger nanoparticles.

The initial samples of photo-thermo-refractive glass demonstrated luminescence under 340 nm excitation originating from the set of glass dopants. Normalized luminescence spectra of Ag.01, Ag.025, and Ag.05 samples are shown in Figure 2a. It can be seen that Ag.01 and

Ag.025 glasses had very similar luminescence, consisting of a single band corresponding to the emission of Ce^{3+} [35,36]. The fluorescence decay curves of Ag.01 and Ag.025 samples were fitted with a tri-exponential function (Figure 2b); the corresponding average fluorescence lifetimes were 23.7 and 28.3 ns. The Ag.05 luminescence spectrum differed significantly from those of the Ag.01 and Ag.025 glasses; an increase in the Ag_2O concentration led to the appearance of a broad luminescence signal in the range of 425–700 nm (Figure 2a). Time-resolved spectroscopy revealed that the additional emission from the Ag.05 sample was mostly long-lived phosphorescence with a maximum at 510 nm and a lifetime of 100 μs (Figure 2c). Comparison of the short-lived emission spectra of Ag.05 and Ag.01 shows that the fluorescence of Ag.05 was wider and tailed more to the longer wavelength region (Figure 2d). The combination of fluorescence and phosphorescence with emission in blue and green-yellow parts of the spectra is a characteristic feature of the luminescence of silver clusters in the studied glass matrix [24,37]. Additionally, we observed a decrease in the fluorescence lifetime to 16.9 ns, which may originate from the contribution of silver clusters, since the fluorescence lifetime of silver clusters in the studied glass varies from 3.2 to 4.5 ns. Therefore, the formation of luminescent silver species together with plasmonic nanoparticles was observed after the synthesis and annealing of Ag.05 glass.

Table 1. Size of silver nanoparticles calculated via Mie theory.

Heat Treatment Temperature, °C	Ag.01	Ag.025	Ag.05
Initial glass	-	-	6 nm
350	-	-	6 nm
400	1 nm	-	6 nm
450	2 nm	4 nm	6 nm

The initial samples of Ag.01, Ag.025, and Ag.05 glasses had a relatively low luminescence quantum yield ranging from 0.06 to 0.1, which decreased by a factor of 2–5 to 0.02–0.05 after UV irradiation (Table 2). The decrease in the luminescence quantum efficiency after UV irradiation corresponded to the transition of Ce^{3+} to Ce^{4+}, which is generally non-luminescent [38]. UV irradiation of the initial glass samples also changed the shape of the luminescence spectra. Comparison of Figure 2a,e shows that the contribution of the Ce^{3+} emission peak to the glass luminescence was minimized with the increasing concentration of Ag_2O in the glass; the simultaneous appearance of a broad emission band with a maximum at 580 nm was observed. This emission band should belong to some transition species in the glass formed under the UV irradiation, rather than to luminescent clusters. The maximum of silver clusters' phosphorescence was located at shorter wavelengths near 540–550 nm (Figure 3). Also, the broad emission band at 580 nm decayed mainly during the time of excitation flash from the xenon lamp; phosphorescence was only a small fraction of the total emission from the UV-irradiated Ag.025 sample (Figure 2f). Additionally, it was assumed earlier that the main electron acceptors at the stage of UV irradiation of photo-thermo-refractive glass are Sb^{5+} ions, which form $(Sb^{5+})^-$ centers and donate the trapped electrons to Ag^+ ions only during further heat treatment [39].

The UV-irradiated samples of Ag.01, Ag.025, and Ag.05 glasses were heat treated at 350, 400, and 450 °C for 3 h to initiate the intense formation of silver clusters. After heat treatment, the glass samples demonstrated bright white emission under UV excitation (Figure 4a). The quantum yield of luminescence increased from 0.02–0.05 to a maximum value of 0.43 for the Ag.05 sample after heat treatment at 400 °C. After heat treatment, the luminescence spectra of the glass samples demonstrated continuous emission from 400 to 720 nm (Figure 3), except the Ag.01 glass after the heat treatment at 450 °C, for which luminesce was mainly in the red part of the spectrum (Figure 3c). Using time-resolved spectroscopy, the emission of glass samples was separated into fluorescence and phosphorescence with lifetimes of several nanoseconds and hundreds of microseconds, respectively. The coexistence of fluorescence and phosphorescence is a specific feature of the silver cluster emission in glasses and was observed earlier for a similar glass system with silver clusters formed using the Na^+-Ag^+ ion exchange technique [40]. It can be seen that, for all concentrations of Ag_2O in glass, an increase in the heat treatment temperature led to a decrease in the fluorescence contribution to the total emission spectra. This tendency partly originates from the reabsorption of cluster emission, but cannot be totally attributed to the inner-filter effect, considering that the absorption spectra of Ag.025 glass heat treated at 350 and 400 °C were almost identical (Figure 1c), while the contribution of fluorescence lowered from 37% to 28% of the total light emission. Additionally, only for Ag.05 glass samples with the most prominent plasmon peak of silver nanoparticles, we observed a clear reabsorption mark at 410 nm (Figure 3g–i). Therefore, changes in the fluorescence to phosphorescence ratio should originate from the formation of different sets of luminescent silver clusters after heat treatment at different temperatures.

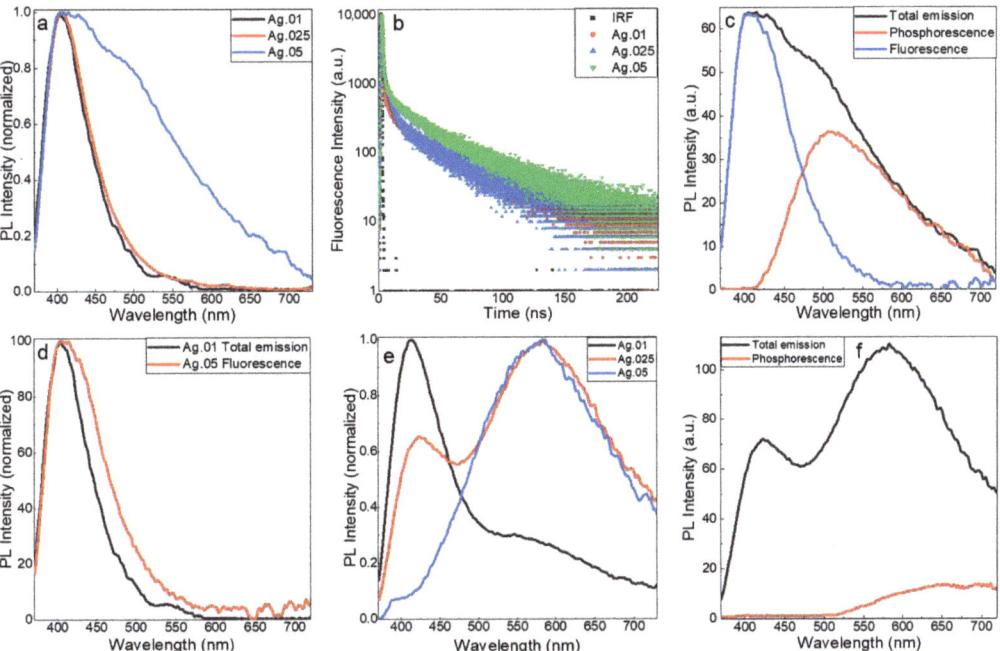

Figure 2. Normalized luminescence spectra of the initial glass samples, λ_{ex} = 340 nm (**a**). Fluorescence decay curves of the initial glass samples registered at 405 nm, λ_{ex} = 340 nm (**b**). Fluorescence, phosphorescence, and total emission spectra of the initial Ag.05 glass sample, λ_{ex} = 340 nm (**c**). Comparison of the Ag.01 and Ag.05 glass fluorescence spectra, λ_{ex} = 340 nm (**d**). Normalized luminescence spectra of the glass samples after UV irradiation, λ_{ex} = 340 nm (**e**). Phosphorescence and total emission spectra of the Ag.025 glass after UV irradiation, λ_{ex} = 340 nm (**f**).

Table 2. Quantum yields of Ag.01, Ag.025, and Ag.05 glass samples under 340 nm excitation.

Heat Treatment Temperature, °C	Ag.01	Ag.025	Ag.05
Initial glass	0.1	0.06	0.09
UV-irradiated initial glass	0.02	0.03	0.05
350	0.25	0.35	0.34
400	0.16	0.42	0.43
450	0.07	0.32	0.3

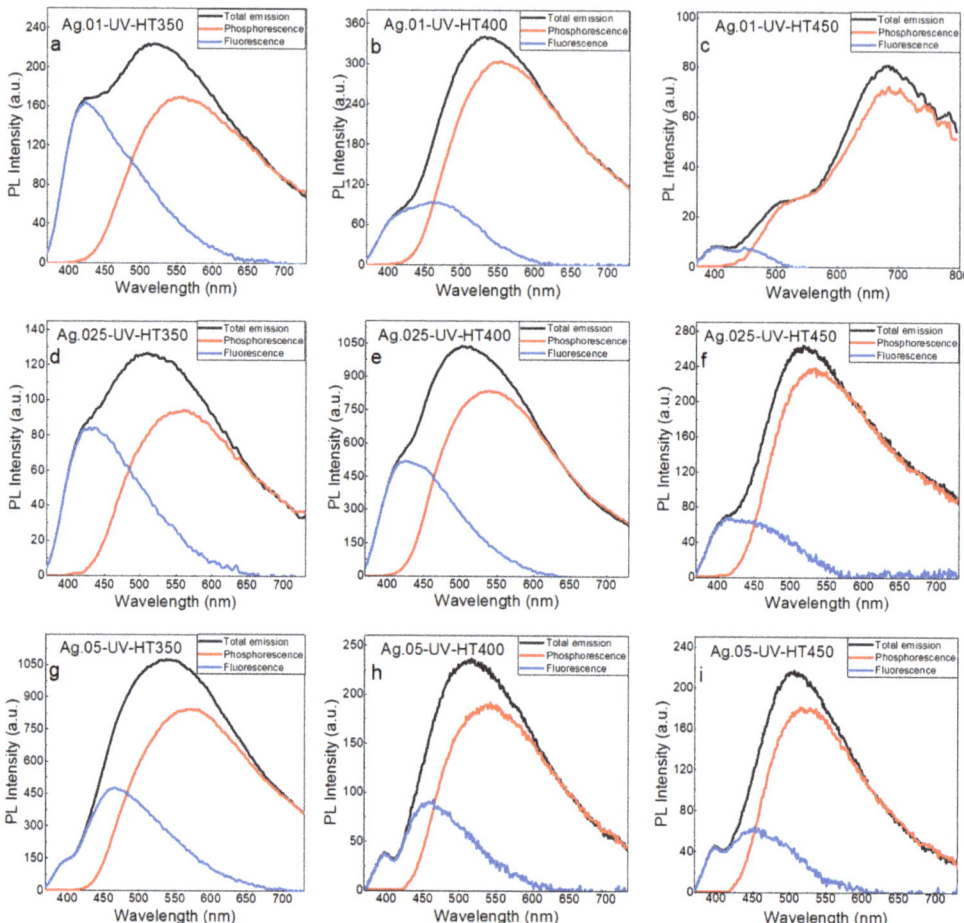

Figure 3. Fluorescence, phosphorescence, and total emission spectra of Ag.01 glass after UV irradiation and heat treatment at (**a**) 350, (**b**) 400, and (**c**) 450 °C. Fluorescence, phosphorescence, and total emission spectra of Ag.025 glass after UV irradiation and heat treatment at (**d**) 350, (**e**) 400, and (**f**) 450 °C. Fluorescence, phosphorescence, and total emission spectra of Ag.05 glass after UV irradiation and heat treatment at (**g**) 350, (**h**) 400, and (**i**) 450 °C. All spectra were measured with λ_{ex} = 340 nm.

Although the heat-treated samples of Ag.05 glass manifested clear plasmon absorption bands, the emission of silver clusters in these samples was not quenched. The photoluminescence quantum yield of the Ag.05 sample heat treated at 400 °C reached the maximum observed value of 0.43 (Table 2). This result indicates that luminescent silver clusters and plasmonic nanoparticles can coexist in glass without negative effects on the emissive properties of the clusters. It has been shown that a characteristic emission of silver clusters was observed for Ag.05 glass right after synthesis (Figure 2a); further development of the effective cluster luminescence after UV irradiation and heat treatment implies that a significant part of silver in glass did not form plasmonic nanoparticles and remained in the form of Ag^+ ions. The fluorescence and phosphorescence lifetimes (Table 3), as well as the spectral properties (Figure 3), were the same for all heat-treated samples, except for the Ag.01 sample after heat treatment at 450 °C. Average fluorescence lifetimes coincided within 0.5 ns, and phosphorescence lifetimes had values of 120–141 µs. Unlike the lifetimes of emission, quantum yields of luminescence differed by more than a factor of two, from

0.16 to 0.43. Considering that the applied method of quantum yield measurements determines the external quantum efficiency, this significant difference should originate from the existence of passive absorption at the excitation wavelength. Possible sources of passive absorption are plasmonic nanoparticles and other non-luminescent silver species, as well as the residual Ce^{3+} ions. To evaluate the possible application of the studied glass samples for the generation of white light, CIE chromaticity coordinates were calculated for the selected glass samples (Figure 4b).

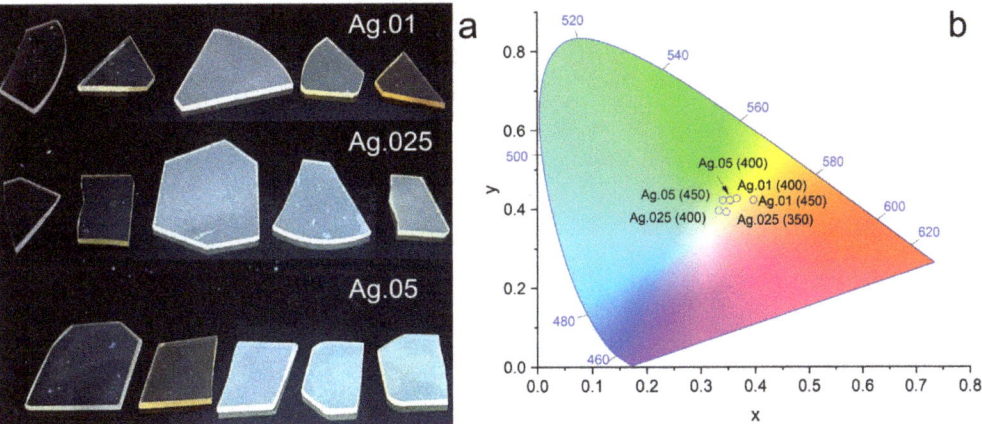

Figure 4. Photo of initial, UV-irradiated, and heat-treated glass samples at temperatures 350, 400, and 450 °C (from left to right) under UV illumination (**a**). CIE chromaticity diagram of the selected glass samples. Temperatures of heat treatment are presented in brackets (**b**).

Table 3. Fluorescence and phosphorescence lifetimes of Ag.01, Ag.025, and Ag.05 glass samples under 340 nm excitation. τ_{f_avg}—average fluorescence lifetime registered at λ_{em} = 405 nm for initial glasses and λ_{em} = 400 nm for heat-treated glasses, τ_p—phosphorescence lifetime measured at λ_{em} = 485 nm for initial Ag.05 glass and λ_{em} = 515 nm for heat-treated glasses.

Heat Treatment Temperature, °C	Ag.01		Ag.025		Ag.05	
	τ_{f_avg} (ns)	τ_p (µs)	τ_{f_avg} (ns)	τ_p (µs)	τ_{f_avg} (ns)	τ_p (µs)
Initial glass	23.7	-	28.3	-	16.9	100
350	3.7	141	3.2	132	3.3	135
400	3.6	136	3.4	136	3.4	133
450	4.5	105	3.2	123	3.4	120

Based on the spectroscopic characterization of the synthesized glass samples, we have chosen Ag.01 glass after heat treatment at 350 °C and Ag.025 glass after heat treatment at 400 °C for the analysis of photophysical process rates. The chosen samples had the highest quantum yields of luminescence for a given concentration of Ag_2O and demonstrated no spectral features of plasmonic nanoparticles. The rate constants of cluster fluorescence, phosphorescence, and intersystem crossing were estimated on the basis of the earlier proposed approach [40], which is summarized below. It was supposed that a set of silver clusters formed in the glass is responsible for the bright white luminescence under near-UV excitation. Broad emission spectra originate from inhomogeneous broadening by different local environments in the glass matrix. Based on the observed fluorescence and phosphorescence emission components, a three-level energy diagram with the ground singlet state S_0, excited singlet S_1, and triplet T_1 states (Figure 5a) was applied to analyze the photophysics of silver clusters in glass. After photoexcitation of a silver cluster, the S_1 state is deactivated radiatively via fluorescence with the process rate constant k_f or nonradiatively via intersystem crossing with the process rate constant k_{isc}. The probability of direct S_1 deactivation via internal conversion is negligible, since the energy of the fluorescence transition is significantly higher than the typical phonon energy of silicate glass [41]. After the intersystem crossing, the excited triplet state T_1 is deactivated radiatively via phosphorescence with the process rate constant k_p or nonradiatively with the rate constant k_{nrT}. Similar three-level energy diagrams were used to describe the properties of silver clusters in other glass systems [42,43]. Within the used approach, it was necessary to measure the quantum yield of emission, separate the spectra of fluorescence and phosphorescence, and the emission decays of fluorescence and phosphorescence to estimate the rate constants of silver clusters. It is important to note that the applied method of photophysical process rate determination provides average values of the rate constants, while the properties of a single silver cluster may deviate from them.

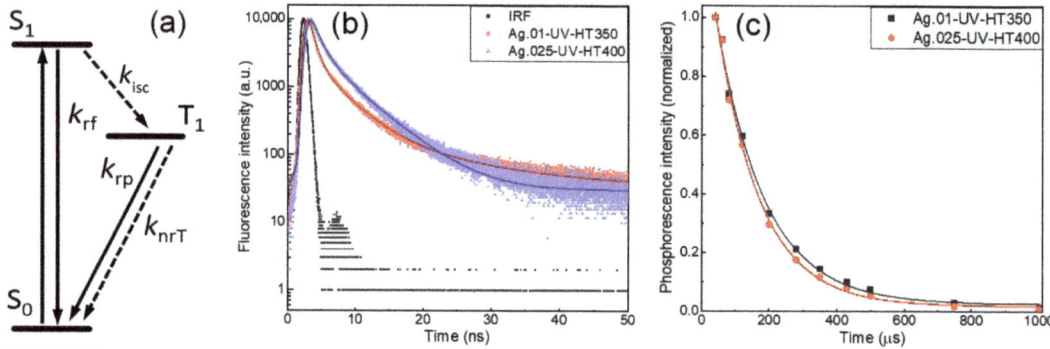

Figure 5. Energy level diagram of silver clusters. k_{rf} and k_{rp}—radiative rate constants of fluorescence and phosphorescence, k_{isc}—intersystem crossing rate constant, k_{nrT}—nonradiative deactivation of the triplet state rate constant (**a**). Fluorescence decay curves of the Ag.01 glass after UV irradiation and heat treatment at 350 °C and Ag.05 glass after UV irradiation and heat treatment at 400 °C measured at λ_{em} = 400 nm, λ_{ex} = 340 nm (**b**). Phosphorescence decay curves of the Ag.01 glass after UV irradiation and heat treatment at 350 °C and Ag.05 glass after UV irradiation and heat treatment at 400 °C measured at λ_{em} = 515 nm, λ_{ex} = 340 nm (**c**).

Figure 5b,c demonstrate the decay curves of fluorescence and phosphorescence. The corresponding lifetimes used for estimation of the photophysical parameters of silver clusters and luminescence quantum yields are summarized in Table 4. Comparison of the luminescence quantum yields and emission lifetimes obtained for silver clusters in the bulk of glass with earlier published results on silver clusters formed by ion exchange in the same glass matrix shows that no significant difference between these synthesis approaches is observed. The most prominent difference in the value of luminescence quantum yield partly originates from the passive absorption at the excitation wavelength since the formation of clusters in ion-exchange layers does not require the cerium in glass composition. The ion exchange procedure used to form silver clusters in photo-thermo-refractive glass provides an average Ag_2O concentration of 5.2 mol.% in a 12.4 µm layer with a surface concentration of 12.7 mol.% [24]. Therefore, the concentrations of silver ions used to form luminescent silver clusters in the glass bulk were 200–500 times lower than in the ion exchange method. Nevertheless, those great differences in the concentration of Ag^+ ions in glass do not lead to the formation of fundamentally different emissive silver clusters. Analysis of the rate constants of photophysical processes shows that a common feature of silver clusters in photo-thermo-refractive glass is a very fast intersystem crossing, leading to the prevalence of phosphorescence in the emission (Table 5). The radiative rate constants of fluorescence and phosphorescence are of the same order of magnitude, as well as the constant rates of the triplet state deactivation.

Table 4. Photophysical parameters of silver clusters in Ag.01 glass after UV irradiation and heat treatment at 350 °C and Ag.05 glass after UV irradiation and heat treatment at 400 °C. τ_i—lifetime of i fluorescence component, τ_{f_avg}—average fluorescence lifetime registered at λ_{em} = 400 nm, τ_p—phosphorescence lifetime measured at λ_{em} = 515 nm. All lifetimes were measured with λ_{ex} = 340 nm. Φ_{lum}—luminescence quantum yield of glass samples under 340 nm excitation.

Sample	τ_1, ns	τ_2, ns	τ_3, ns	τ_{f_avg}, ns	τ_p, µs	Φ_{lum}	Reference
Ag.01-UV-HT350	0.8	3.8	14.5	3.7	141	0.25	this work
Ag.025-UV-HT400	1.3	5.1	-	3.4	136	0.42	this work
Ion-exchanged sample	1.5	5.0	-	3.8	110	0.66	[40]

Table 5. Photophysical parameters of silver clusters in Ag.01 glass after UV irradiation and heat treatment at 350 °C and Ag.05 glass after UV irradiation and heat treatment at 400 °C. k_{rf} and k_{rp}—radiative rate constants of fluorescence and phosphorescence, k_{isc}—intersystem crossing rate constant, k_{nrT}—nonradiative deactivation of the triplet state rate constant, Φ_f, Φ_{isc}, and Φ_p—fluorescence, intersystem crossing, and phosphorescence quantum yields.

Sample	k_{rf} (s^{-1})	k_{isc} (s^{-1})	k_{rp} (s^{-1})	k_{nrT} (s^{-1})	Φ_f	Φ_{isc}	Φ_p	Reference
Ag.01-UV-HT350	1.6·10^7	2.5·10^8	1.4·10^3	5.6·10^3	0.06	0.94	0.20	this work
Ag.025-UV-HT400	2.3·10^7	2.7·10^8	2.7·10^3	4.6·10^3	0.08	0.92	0.37	this work
Ion-exchanged sample	3.2·10^7	2.3·10^8	5.6·10^3	3.6·10^3	0.12	0.88	0.61	[40]

4. Conclusions

Luminescent silver clusters were formed in the bulk of photo-thermo-refractive glass containing 0.01, 0.025, and 0.05 mol.% of Ag$_2$O. Spontaneous formation of plasmonic nanoparticles during glass synthesis was observed at 0.05 mol.% of Ag$_2$O in the glass composition. After UV irradiation and heat treatment of the synthesized glasses, efficient formation of silver clusters with bright white luminescence was observed even for the glass samples with plasmonic nanoparticles. The spectroscopic characterization using steady-state and time-resolved techniques revealed that silver clusters in the glass bulk have the same characteristic features as the previously studied silver clusters in ion-exchange layers. The emission spectra of silver clusters consist of fluorescence with an average lifetime of 3.2–4.5 ns and phosphorescence with a lifetime of 105–141 µs. The quantum yield of the studied glass samples varies from 0.25 to 0.42. Based on experimental studies, the estimation of the radiative rate constants of fluorescence and phosphorescence, the rate constant of intersystem crossing, and the rate constant of nonradiative deactivation of the triplet state was performed. The results obtained on fluorescence and phosphorescence lifetimes and quantum yields, as well as the rates of photophysical processes, were compared with the previously obtained results for silver clusters formed in ion-exchange layers of the same glass matrix. It was revealed that different synthesis approaches form close sets of emissive silver clusters, despite the difference in Ag$^+$ concentration by 200–500 times. This result suggests that silver clusters of one type are responsible for the white emission of the studied glasses.

Author Contributions: Formal analysis, L.Y.M.; Investigation, L.Y.M., D.V.M., M.D.S., K.S.Z., A.A.S. and I.E.K.; Writing—original draft, L.Y.M.; Writing—review and editing, L.Y.M. All authors have read and agreed to the published version of the manuscript.

Funding: This research was funded by the Russian Science Foundation, project NO 22-73-10055.

Institutional Review Board Statement: Not applicable.

Informed Consent Statement: Not applicable.

Data Availability Statement: The data presented in this study are available on request from the corresponding author.

Acknowledgments: Fluorescence decay measurements were carried out in the 'Center for Optical and Laser materials research' (Saint Petersburg State University).

Conflicts of Interest: The authors declare no conflict of interest.

References

1. Kuznetsov, A.S.; Tikhomirov, V.K.; Shestakov, M.V.; Moshchalkov, V.V. Ag Nanocluster Functionalized Glasses for Efficient Photonic Conversion in Light Sources, Solar Cells and Flexible Screen Monitors. *Nanoscale* **2013**, *5*, 10065–10075. [CrossRef] [PubMed]
2. Tan, D.; Jiang, P.; Xu, B.; Qiu, J. Single-Pulse-Induced Ultrafast Spatial Clustering of Metal in Glass: Fine Tunability and Application. *Adv. Photonics Res.* **2021**, *2*, 2000121. [CrossRef]
3. Wu, Y.; Lin, H.; Li, R.; Lin, S.; Wu, C.; Huang, Q.; Xu, J.; Cheng, Y.; Wang, Y. Laser-Direct-Writing of Molecule-like Ag_m^{x+} nanoclusters in Transparent Tellurite Glass for 3D Volumetric Optical Storage. *Nanoscale* **2021**, *13*, 19663–19670. [CrossRef] [PubMed]
4. De Cremer, G.; Sels, B.F.; Hotta, J.I.; Roeffaers, M.B.J.; Bartholomeeusen, E.; Coutiño-Gonzalez, E.; Valtchev, V.; De Vos, D.E.; Vosch, T.; Hofkens, J. Optical Encoding of Silver Zeolite Microcarriers. *Adv. Mater.* **2010**, *22*, 957–960. [CrossRef]
5. Dong, X.Y.; Si, Y.; Yang, J.S.; Zhang, C.; Han, Z.; Luo, P.; Wang, Z.Y.; Zang, S.Q.; Mak, T.C.W. Ligand Engineering to Achieve Enhanced Ratiometric Oxygen Sensing in a Silver Cluster-Based Metal-Organic Framework. *Nat. Commun.* **2020**, *11*, 3678. [CrossRef]
6. Qian, S.; Wang, Z.; Zuo, Z.; Wang, X.; Wang, Q.; Yuan, X. Engineering Luminescent Metal Nanoclusters for Sensing Applications. *Coord. Chem. Rev.* **2022**, *451*, 214268. [CrossRef]
7. Zheng, W.; Li, P.; Wang, C.; Qiao, X.; Qian, G.; Fan, X. Tuning Ag Quantum Clusters in Glass as an Efficient Spectral Converter: From Fundamental to Applicable. *J. Non. Cryst. Solids* **2023**, *599*, 121910. [CrossRef]
8. Fares, H.; Santos, S.N.C.; Santos, M.V.; Franco, D.F.; Souza, A.E.; Manzani, D.; Mendonça, C.R.; Nalin, M. Highly Luminescent Silver Nanocluster-Doped Fluorophosphate Glasses for Microfabrication of 3D Waveguides. *RSC Adv.* **2017**, *7*, 55935–55944. [CrossRef]
9. Aslani, M.; Talebi, R.; Vashaee, D. Coupling Light in Ion-Exchanged Waveguides by Silver Nanoparticle-Based Nanogratings: Manipulating the Refractive Index of Waveguides. *ACS Appl. Nano Mater.* **2022**, *5*, 5439–5447. [CrossRef]
10. de Castro, T.; Fares, H.; Khalil, A.A.; Laberdesque, R.; Petit, Y.; Strutinski, C.; Danto, S.; Jubera, V.; Ribeiro, S.J.L.; Nalin, M.; et al. Femtosecond Laser Micro-Patterning of Optical Properties and Functionalities in Novel Photosensitive Silver-Containing Fluorophosphate Glasses. *J. Non. Cryst. Solids* **2019**, *517*, 51–56. [CrossRef]
11. Sholom, S.; McKeever, S.W.S. Silver Molecular Clusters and the Properties of Radiophotoluminescence of Alkali-Phosphate Glasses at High Dose. *Radiat. Meas.* **2023**, *163*, 106924. [CrossRef]
12. McKeever, S.W.S.; Sholom, S.; Shrestha, N.; Klein, D.M. Build-up of Radiophotoluminescence (RPL) in Ag-Doped Phosphate Glass in Real-Time Both during and after Exposure to Ionizing Radiation: A Proposed Model. *Radiat. Meas.* **2020**, *132*, 106246. [CrossRef]
13. Han, Z.; Dong, X.Y.; Luo, P.; Li, S.; Wang, Z.Y.; Zang, S.Q.; Mak, T.C.W. Ultrastable Atomically Precise Chiral Silver Clusters with More than 95% Quantum Efficiency. *Sci. Adv.* **2020**, *6*, eaay0107. [CrossRef]
14. Cho, J.; Park, J.H.; Kim, J.K.; Schubert, E.F. White Light-Emitting Diodes: History, Progress, and Future. *Laser Photonics Rev.* **2017**, *11*, 1600147. [CrossRef]
15. Grandjean, D.; Coutiño-Gonzalez, E.; Cuong, N.T.; Fron, E.; Baekelant, W.; Aghakhani, S.; Schlexer, P.; D'Acapito, F.; Banerjee, D.; Roeffaers, M.B.J.; et al. Origin of the Bright Photoluminescence of Few-Atom Silver Clusters Confined in LTA Zeolites. *Science* **2018**, *361*, 686–690. [CrossRef]
16. Coutino-Gonzalez, E.; Baekelant, W.; Grandjean, D.; Roeffaers, M.B.J.; Fron, E.; Aghakhani, M.S.; Bovet, N.; Van Der Auweraer, M.; Lievens, P.; Vosch, T.; et al. Thermally Activated LTA(Li)-Ag Zeolites with Water-Responsive Photoluminescence Properties. *J. Mater. Chem. C* **2015**, *3*, 11857–11867. [CrossRef]
17. Baekelant, W.; Aghakhani, S.; Fron, E.; Martin, C.; Woong-Kim, C.; Steele, J.A.; De Baerdemaeker, T.; D'Acapito, F.; Chernysov, D.; Van Der Auweraer, M.; et al. Luminescent Silver-Lithium-Zeolite Phosphors for near-Ultraviolet LED Applications. *J. Mater. Chem. C* **2019**, *7*, 14366–14374. [CrossRef]

18. Fenwick, O.; Coutiño-Gonzalez, E.; Grandjean, D.; Baekelant, W.; Richard, F.; Bonacchi, S.; De Vos, D.; Lievens, P.; Roeffaers, M.; Hofkens, J.; et al. Tuning the Energetics and Tailoring the Optical Properties of Silver Clusters Confined in Zeolites. *Nat. Mater.* **2016**, *15*, 1017–1022. [CrossRef]
19. Coutino-Gonzalez, E.; Roeffaers, M.B.J.; Dieu, B.; De Cremer, G.; Leyre, S.; Hanselaer, P.; Fyen, W.; Sels, B.; Hofkens, J. Determination and Optimization of the Luminescence External Quantum Efficiency of Silver-Clusters Zeolite Composites. *J. Phys. Chem. C* **2013**, *117*, 6998–7004. [CrossRef]
20. Kuznetsov, A.S.; Tikhomirov, V.K.; Moshchalkov, V.V. UV-Driven Efficient White Light Generation by Ag Nanoclusters Dispersed in Glass Host. *Mater. Lett.* **2013**, *92*, 4–6. [CrossRef]
21. Hu, T.; Zheng, W.; Liu, Z.; Jia, J.; Xu, X.; Xu, Q.; Qiao, X.; Fan, X. Strategies to Host Silver Quantum Clusters in Borosilicate Glass: How to Mutually Fulfill PL Efficiency and Chemical Stability? *J. Non-Crystalline Solids X* **2022**, *16*, 100132. [CrossRef]
22. Sgibnev, Y.M.; Nikonorov, N.V.; Ignatiev, A.I. High Efficient Luminescence of Silver Clusters in Ion-Exchanged Antimony-Doped Photo-Thermo-Refractive Glasses: Influence of Antimony Content and Heat Treatment Parameters. *J. Lumin.* **2017**, *188*, 172–179. [CrossRef]
23. Dubrovin, V.D.; Ignatiev, A.I.; Nikonorov, N.V.; Sidorov, A.I.; Shakhverdov, T.A.; Agafonova, D.S. Luminescence of Silver Molecular Clusters in Photo-Thermo-Refractive Glasses. *Opt. Mater.* **2014**, *36*, 753–759. [CrossRef]
24. Mironov, L.Y.; Marasanov, D.V.; Ulshina, M.D.; Sgibnev, Y.M.; Kolesnikov, I.E.; Nikonorov, N.V. The Role of Thermally Activated Quenching and Energy Migration in Luminescence of Silver Clusters in Glasses. *J. Phys. Chem. C* **2022**, *126*, 13863–13869. [CrossRef]
25. West, B.R. Ion-Exchanged Glass Waveguide Technology: A Review. *Opt. Eng.* **2011**, *50*, 071107. [CrossRef]
26. Stepanov, A.L.; Hole, D.E.; Townsend, P.D. Formation of Silver Nanoparticles in Soda-Lime Silicate Glass by Ion Implantation near Room Temperature. *J. Non. Cryst. Solids* **1999**, *260*, 65–74. [CrossRef]
27. Arnold, G.W.; Borders, J.A. Aggregation and Migration of Ion-implanted Silver in Lithia-alumina-silica Glass. *J. Appl. Phys.* **1977**, *48*, 1488–1496. [CrossRef]
28. Lumeau, J.; Zanotto, E.D. A Review of the Photo-Thermal Mechanism and Crystallization of Photo-Thermo-Refractive (PTR) Glass. *Int. Mater. Rev.* **2017**, *62*, 348–366. [CrossRef]
29. Ivanov, S.; Dubrovin, V.; Nikonorov, N.; Stolyarchuk, M.; Ignatiev, A. Origin of Refractive Index Change in Photo-Thermo-Refractive Glass. *J. Non. Cryst. Solids* **2019**, *521*, 119496. [CrossRef]
30. Efimov, A.M.; Ignatiev, A.I.; Nikonorov, N.V.; Postnikov, E.S. Quantitative UV–VIS Spectroscopic Studies of Photo-Thermo-Refractive Glasses. II. Manifestations of Ce^{3+} and Ce(IV) Valence States in the UV Absorption Spectrum of Cerium-Doped Photo-Thermo-Refractive Matrix Glasses. *J. Non. Cryst. Solids* **2013**, *361*, 26–37. [CrossRef]
31. Paul, A.; Mulholland, M.; Zaman, M.S. Ultraviolet Absorption of Cerium(III) and Cerium(IV) in Some Simple Glasses. *J. Mater. Sci.* **1976**, *11*, 2082–2086. [CrossRef]
32. Hövel, H.; Fritz, S.; Hilger, A.; Kreibig, U.; Vollmer, M. Width of Cluster Plasmon Resonances: Bulk Dielectric Functions and Chemical Interface Damping. *Phys. Rev. B* **1993**, *48*, 18178–18188. [CrossRef]
33. Arnold, G.W. Near-surface Nucleation and Crystallization of an Ion-implanted Lithia-alumina-silica Glass. *J. Appl. Phys.* **1975**, *46*, 4466–4473. [CrossRef]
34. Jiménez, J.A.; Sendova, M.; Liu, H. Evolution of the Optical Properties of a Silver-Doped Phosphate Glass during Thermal Treatment. *J. Lumin.* **2011**, *131*, 535–538. [CrossRef]
35. Sontakke, A.D.; Ueda, J.; Tanabe, S. Effect of Synthesis Conditions on Ce^{3+} Luminescence in Borate Glasses. *J. Non. Cryst. Solids* **2016**, *431*, 150–153. [CrossRef]
36. Chewpraditkul, W.; Shen, Y.; Chen, D.; Yu, B.; Prusa, P.; Nikl, M.; Beitlerova, A.; Wanarak, C. Luminescence and Scintillation of Ce^{3+}-Doped High Silica Glass. *Opt. Mater.* **2012**, *34*, 1762–1766. [CrossRef]
37. Mironov, L.Y.; Marasanov, D.V.; Sgibnev, Y.M.; Sannikova, M.D.; Kulpina, E. V Influence of Reducing Agent Concentration on the Luminescence and Photophysical Processes Constant Rates of Silver Clusters in Silica-Based Glass. *J. Lumin.* **2023**, *261*, 119918. [CrossRef]
38. He, Y.; Pei, G.; Liu, J.; Fang, Z.; Wang, L.; Jiang, S.; Yu, B.; Gou, J.; Liu, S.F. Utilizing the Energy Transfer of Ce^{4+}- and Ce^{3+}-Tb^{3+} to Boost the Luminescence Quantum Efficiency up to 100% in Borate Glass. *J. Phys. Chem. C* **2022**, *126*, 5838–5846. [CrossRef]
39. Efimov, A.M.; Ignatiev, A.I.; Nikonorov, N.V.; Postnikov, E.S. Quantitative UV-VIS Spectroscopic Studies of Photo-Thermo-Refractive Glasses. I. Intrinsic, Bromine-Related, and Impurity-Related UV Absorption in Photo-Thermo-Refractive Glass Matrices. *J. Non. Cryst. Solids* **2011**, *357*, 3500–3512. [CrossRef]
40. Marasanov, D.V.; Mironov, L.Y.; Sgibnev, Y.M.; Kolesnikov, I.E.; Nikonorov, N.V. Luminescence and Energy Transfer Mechanisms in Photo-Thermo-Refractive Glasses Co-Doped with Silver Molecular Clusters and Eu^{3+}. *Phys. Chem. Chem. Phys.* **2020**, *22*, 23342–23350. [CrossRef]
41. Cao, R.; Lu, Y.; Tian, Y.; Huang, F.; Guo, Y.; Xu, S.; Zhang, J. 2 Mm Emission Properties and Nonresonant Energy Transfer of Er^{3+} and Ho^{3+} Codoped Silicate Glasses. *Sci. Rep.* **2016**, *6*, 37873. [CrossRef] [PubMed]

42. Zheng, W.; Zhou, B.; Ren, Z.; Xu, X.; Yang, G.; Qiao, X.; Yan, D.; Qian, G.; Fan, X. Fluorescence–Phosphorescence Manipulation and Atom Probe Observation of Fully Inorganic Silver Quantum Clusters: Imitating from and Behaving beyond Organic Hosts. *Adv. Opt. Mater.* **2022**, *10*, 2101632. [CrossRef]
43. Li, L.; Yang, Y.; Zhou, D.; Yang, Z.; Xu, X.; Qiu, J. Influence of the Eu^{2+} on the Silver Aggregates Formation in Ag^+-Na^+ Ion-Exchanged Eu^{3+}-Doped Sodium-Aluminosilicate Glasses. *J. Am. Ceram. Soc.* **2014**, *97*, 1110–1114. [CrossRef]

Disclaimer/Publisher's Note: The statements, opinions and data contained in all publications are solely those of the individual author(s) and contributor(s) and not of MDPI and/or the editor(s). MDPI and/or the editor(s) disclaim responsibility for any injury to people or property resulting from any ideas, methods, instructions or products referred to in the content.

Article

New Glasses in the PbCl$_2$–PbO–B$_2$O$_3$ System: Structure and Optical Properties

Dmitry Butenkov [1], Anna Bakaeva [1], Kristina Runina [1], Igor Krol [1], Maria Uslamina [2], Aleksandr Pynenkov [2], Olga Petrova [1] and Igor Avetissov [1,*]

[1] Department of Chemistry and Technology of Crystals, D. Mendeleev University of Chemical Technology of Russia (MUCTR), Moscow 125480, Russia; dabutenkov@gmail.com (D.B.); bakaevanna@mail.ru (A.B.); runinakristina@mail.ru (K.R.); krol_2.0@mail.ru (I.K.); petrova.o.b@muctr.ru (O.P.)

[2] Institute of High Technologies and New Materials, National Research Mordovia State University, Saransk 430005, Russia; uslaminam@mail.ru (M.U.); alekspyn@yandex.ru (A.P.)

* Correspondence: avetisov.i.k@muctr.ru

Abstract: New oxychloride lead borate glasses in the xPbCl$_2$–(50-0.5x)PbO–(50-0.5x)B$_2$O$_3$ system were synthesized with a maximum lead chloride content of 40 mol%. The characteristic temperatures and mechanical and optical properties were studied. The incorporation of lead chloride led to a significant expansion of the transparency range in the UV (up to 355 nm) and IR regions (up to 4710 nm). Decreases in the Vickers hardness, density, and glass transition temperature were the consequences of a change in the structure. The studied glasses are promising materials for photonics and IR optics. The structure of the PbCl$_2$–PbO–B$_2$O$_3$ system was analyzed in detail using vibrational spectroscopy and X-ray diffraction.

Keywords: oxychloride glasses; lead chloride; optical absorption; glass network structure

Citation: Butenkov, D.; Bakaeva, A.; Runina, K.; Krol, I.; Uslamina, M.; Pynenkov, A.; Petrova, O.; Avetissov, I. New Glasses in the PbCl$_2$–PbO–B$_2$O$_3$ System: Structure and Optical Properties. *Ceramics* **2023**, *6*, 1348–1364. https://doi.org/10.3390/ceramics6030083

Academic Editors: Georgiy Shakhgildyan and Michael I. Ojovan

Received: 26 May 2023
Revised: 22 June 2023
Accepted: 25 June 2023
Published: 27 June 2023

Copyright: © 2023 by the authors. Licensee MDPI, Basel, Switzerland. This article is an open access article distributed under the terms and conditions of the Creative Commons Attribution (CC BY) license (https://creativecommons.org/licenses/by/4.0/).

1. Introduction

The variety of compositions and extremely interesting properties of glasses based on boron oxide have attracted the attention of many researchers. In the last few decades, the importance of borate glasses has increased because they exhibit an exceptional combination of properties, such as transparency in a wide spectral range, mechanical strength, chemical resistance, as well as thermal shock and crystallization resistance [1].

Borate glasses containing lead compounds have unique characteristics. The inclusion of PbO and PbX$_2$ (X = F, Cl, and Br) in glass compositions leads to a significant increase in electrical conductivity [2], transparency in the IR and UV ranges [3,4], high refractive indices [5], and optical nonlinearity [6].

Oxyhalide glasses are very promising as matrix materials for applications such as modern laser systems and fiber communication lines [7]. The incorporation of lead halides (PbX$_2$) into glasses leads to low phonon energies, which makes them promising materials due to a decrease in nonradiative losses because of multiphonon relaxations [8]. The optical properties of glasses of the PbX$_2$–PbO–B$_2$O$_3$ system doped with d-elements [5,9] and f-elements [4,10,11] have been studied. The positive effect of lead halides on their spectral and luminescent properties has been proven. However, the preparation and usage of many oxyhalide glasses are often limited due to their high tendency to crystallize, the toxicity of reagents, and the rather high content of OH$^-$ groups [4,10]. There is no information on the preparation of borate glasses with a PbCl$_2$ content of more than 25 mol% and there are limited data on their optical properties in the IR region and their structures.

This work aimed to develop a synthesis technique for new oxychloride lead borate glasses with a large amount of lead chloride (up to 40 mol%) and to investigate their optical, mechanical, and thermal properties.

2. Materials and Methods

The glasses were synthesized with the general formula xPbCl$_2$-(50-0.5x)PbO-(50-0.5x)B$_2$O$_3$, where x varied from 0 to 40 mol% with a 5 mol% step. PbCl$_2$, PbO powder preparations (99.9, Himkraft, Kaliningrad, Russia) and B$_2$O$_3$ glassy strips (99.9, Reachim, Moscow, Russia) were used as initial components for synthesis. The necessary amounts of components were weighed using an Explorer Ohaus PA64 electronic balance (Ohaus, Shanghai, China). The mixture was homogenized by repeated grinding in an agate mortar. The sample weight for one experiment was ~15 g. The charge was placed in a corundum crucible with a cover and melted for 30 min at 900 °C in a PM-12M1 muffle furnace (EVS, St. Petersburg, Russia) under ambient atmosphere. The melt was then cast into a glassy carbon mold and quickly pressed by another glassy carbon mold. To reduce quenching stresses, the glasses were annealed in a SNOL 7.2/1100 furnace (Umega Group, Ukmergė, Lithuania) at 240 °C for samples with PbCl$_2$ content $\geq$20 mol% and at 300 °C for samples with PbCl$_2$ content $\leq$ 15 mol%. The residual stress was controlled by using the optical polarization method with a PKS-250M polariscope-polarimeter (Zagorsk Optical and Mechanical Plant, Sergiev Posad, Russia). Plane-parallel samples of 2 mm thickness were polished using a T-080.00.00 stone-cutting machine (Togran, Moscow, Russia).

The amorphous nature of the glasses was confirmed by X-ray diffraction analysis using an Equinox 2000 X-ray diffractometer (Thermo Fisher Scientific, Saint-Herblain, France) with CuK$_\alpha$ radiation (λ = 1.54060 Å) in the 2Θ angle range from 10° to 100°, with a 0.01° scanning step and 2 s/step exposure. The diffraction patterns were analyzed using Match! software (2003–2015 CRYSTAL IMPAC T, Bonn, Germany).

We used X-ray fluorescent spectroscopy (XFS) to estimate the element composition of the synthesized glasses. The measurements were carried out using a TESCAN VEGA3-LMU scanning electron microscope (TESCAN ORSAY HOLDING, Brmo, Czech Republic) equipped with an X-MAX-50 EDS detector (Oxford Instruments, Abingdon, UK) at 30 kV voltage.

The absorption spectra were recorded using a JASCO V-770 spectrophotometer (JASCO Corporation, Tokyo, Japan) in the range of 190–2700 nm. The absorption spectra were obtained using a Bruker Tensor 27 FT-IR spectrometer (Bruker, Berlin, Germany) in the 1.25–27.00 µm range.

The refractive index of glass samples was measured on a Metricon 2010 refractometer (Metricon, Pennington, NJ, USA). The measurement method was based on determining the critical angle of incidence at which light began to penetrate the sample volume through the surface of the measuring prism, similar to the Abbe refractometer. The refractive index was measured at three wavelengths: 633, 969, and 1539 nm.

Glass transition (T_g), crystallization (T_c), and melting (T_m) temperatures were determined by differential scanning calorimetry (DSC) using an SDT Q-600 thermal analyzer (TA Instruments, New Castle, DE, USA) at 10 °C/min rate from RT to 800 °C. Aluminum oxide (Al$_2$O$_3$) was used as a reference. For analysis, glass samples were preliminarily ground into powder. At the same time, thermogravimetric analysis was performed to assess the volatilization of glass components.

The structure of the glasses was studied using Fourier transform IR spectra and Raman spectra (RS). FT-IR spectroscopy was performed on a Bruker Tensor 27 spectrometer (Bruker, Berlin, Germany). For FT-IR measurements, the glass samples were ground into a fine powder and then mixed with KBr powder in a weight ratio of 0.01:0.19. The prepared mixture was pressed into pellets, which were immediately measured (in the range of 400–8000 cm^{-1}) to avoid exposure to atmospheric moisture. Raman spectra were recorded in the 80–3500 cm^{-1} range using standard laser excitation on a Vertex 70 FT-IR spectrometer with the Raman module RAM II (Bruker, Berlin, Germany) equipped with a 1064 nm neodymium laser. The laser power was 350 mW. Powder preparations were used for Raman measurements.

The glass density was determined using the hydrostatic weighing method with a MERTECH M-ER 123 ACF JR balance and special equipment (MERTECH, Moscow, Russia).

Vickers hardness of glasses was measured by indenting a diamond pyramid with a square base and an angle between the faces equal to 136°. To increase the statistical reliability, the measurements were carried out 5 times for each of the four weight loads (50, 100, 150, and 200 g), and the measurement results were averaged. For each point, the average deviation was calculated, and these were then averaged. Samples were measured using a PMT-3 tester (LOMO, St. Petersburg, Russia).

3. Results and Discussion

3.1. Glass Composition and Structure Characterization

The compositions of the obtained glasses are shown in Figure 1.

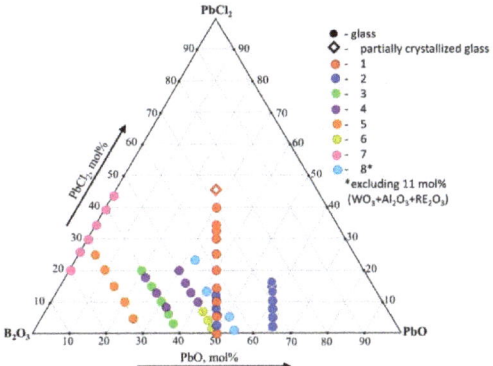

Figure 1. The compositions of glasses in the $PbCl_2$–PbO–B_2O_3 system: 1—our work; 2—[2]; 3—[12]; 4—[13,14]; 5—[5]; 6—[15]; 7—[16]; 8—[4,10,11].

The diffraction patterns of the synthesized glasses confirmed the amorphous nature of the samples (Figure 2).

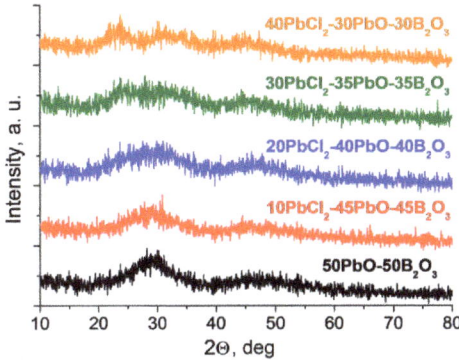

Figure 2. XRD patterns of glasses in the $xPbCl_2$–$(50-0.5x)PbO$–$(50-0.5x)B_2O_3$ system.

Figure 2 shows that the glass halo underwent significant changes upon the addition of $PbCl_2$, which indicated a change in the short-range order structure. We assumed that at high $PbCl_2$ concentrations, the local glass structure tended toward the structure of crystalline $PbCl_2$ but with a greater degree of disorder, which was specific for the glassy state (Figure S1) [17,18]. Heavily doped glasses (>20%) were possibly partly crystallized (nanocrystals formation), but the limited resolution of the patterns did not allow a clear conclusion. When glasses with a $PbCl_2$ content of more than 40% were synthesized, partial crystallization of the melt occurred with the formation of the PbCl$_2$ phase (Figure S1). The PbCl$_2$ phase was rhombic (space group $Pnma$, a = 7.623 Å, b = 9.048 Å, c = 4.535 Å, Z = 4). Card ID 01-084-1177 (MATCH) or pdf 26-1159 PCPDFWIN.

The synthesis of oxychloride glasses is complicated due to the high hygroscopicity of Cl-containing components. In the process of glass melting, in addition to direct volatilization, pyrohydrolysis occurs [4]:

$$PbCl_2 + H_2O \rightarrow PbO + 2HCl\uparrow \quad (1)$$

An exchange reaction is also possible [19]:

$$3PbCl_2 + B_2O_3 \rightarrow 3PbO + 2BCl_3\uparrow \quad (2)$$

The results of the elemental analysis (Table S1) showed that the Cl/Pb ratio in the synthesized glasses remained almost the same as in the initial batch (Figure 3).

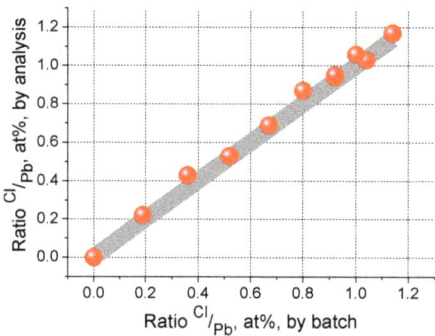

Figure 3. Comparison of Cl/Pb ratio in the batch and glasses synthesized in the xPbCl$_2$–(50-0.5x)PbO–(50-0.5x)B$_2$O$_3$ system.

To evaluate the volatilization of the glass components during synthesis, we applied a direct weighing technique. It was shown that volatilization remained almost constant for all compositions with PbCl$_2$ below 35 mol% (Figure 4). Thus, we assumed that B$_2$O$_3$ and water volatilized simultaneously with PbCl$_2$. However, there was no way to confirm this assumption, since the XFS technique is limited in detecting light elements such as boron (Z_B = 5). This is the reason that the data in Table S1 are presented in the form of ratios of elements.

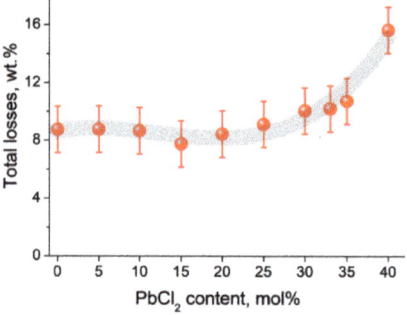

Figure 4. Volatilization of glass components from the melt during synthesis depending on PbCl$_2$ content for glasses in the xPbCl$_2$–(50-0.5x)PbO–(50-0.5x)B$_2$O$_3$ system. The line provides a guide for the eyes.

The synthesized glasses contained a small amount of aluminum (Table S1) because of the dissolution of the corundum crucible by the aggressive PbO-based melt. However, Al_2O_3 has a positive effect on the spectral and mechanical properties of glasses due to it being a network former [20]. We did not observe significant losses of $PbCl_2$ during synthesis, which made it possible to correctly describe the physicochemical properties of the glasses based on their batch composition.

The glass structure was studied by Raman spectroscopy in the 80–3500 cm^{-1} range, which was divided into 3 parts for detailed analysis: 80–250 cm^{-1} (Figures 5, S2 and S3), 250–1750 cm^{-1} (Figure 6), and 1750–3500 cm^{-1} (Figure S4). In Raman spectroscopy, vibrations with wavenumbers less than 400 cm^{-1} are usually attributed to vibrational modes associated with heavy metal atoms. In addition, the frequencies of halide compounds with predominantly ionic bonds are generated in this range [21,22].

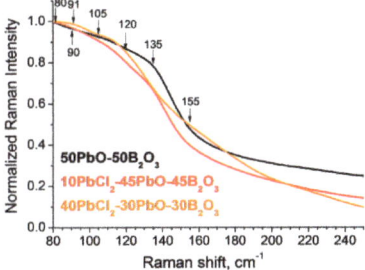

Figure 5. Raman spectra of $xPbCl_2$–$(50-0.5x)PbO$–$(50-0.5x)B_2O_3$ glasses (80–250 cm^{-1} range).

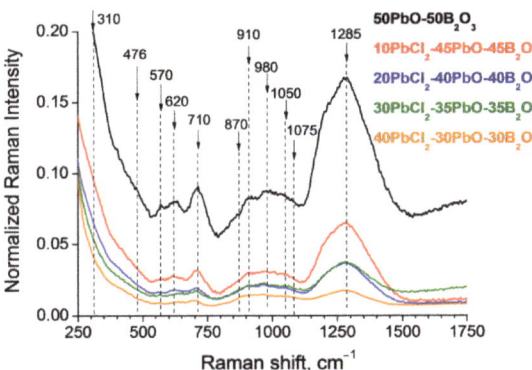

Figure 6. Raman spectra of $xPbCl_2$–$(50-0.5x)PbO$–$(50-0.5x)B_2O_3$ glasses (250–1750 cm^{-1} range).

We observed a broadened peak in the 80–250 cm^{-1} range (Figure 5), which was a superposition of several fundamental modes and a bosonic peak [4,23]. The assignment of vibrational modes to the corresponding structural groupings is presented in Table 1. The observed broadened peak underwent significant changes depending on the glass composition. For the $50PbO$–$50B_2O_3$ glass composition, we observed 4 main modes at 80, 90, 105, and 135 cm^{-1} (Figure S2). These were the result of bending vibrations in PbO_4 structural units [23], stretching of the ionic component of the Pb–O bond [24], Pb^{2+} cations in the glass network [25], and covalent symmetric stretching of the Pb–O bond in PbO_4 tetrahedra [26], respectively. In lead borate glasses with a high PbO content, $[PbO_4]^{2-}$ pyramids are glass network-forming units and most of the lead went into building these pyramids. The rest of the lead acted as a charge compensator in the form of free Pb^{2+} ions. $[PbO_4]^{2-}$ units preferentially bridge bonds with BO_3 groups rather than BO_4 groups [25]. The dominant units in these glasses are BO_3 and $[PbO_4]^{2-}$ with Pb—O—B bridges between them. The introduction of lead chloride caused a sharp decrease in mode intensity around 135 cm^{-1}, indicating the destruction of these bridges [25].

In addition, local peaks with maxima at 80 and 90 cm^{-1} were strongly smoothed and had reduced intensities, which also indicated the disappearance of PbO$_n$ units in the glass network. On the other hand, we observed a generation of new bands at 91, 105, and 120 cm^{-1}. They belonged to vibrations of Cl$^-$ [27] ions, Pb^{2+} [28] cations, and stretching vibrations of the Pb–Cl bond [29]. The further growth of PbCl$_2$ concentration up to 40 mol% resulted in the generation of new modes at 150–180 cm^{-1} [27], while the modes related to the structural units of PbO$_n$ practically disappeared. Cl$^-$ acts as a non-bridging anion and two halogen anions replace one O^{2-}. This leads to the destruction of [PbO$_4$]$^{2-}$ units and the formation of Pb^{2+} cations. The appearance of Raman bands related to vibrations of chlorine anions and the Pb–Cl bond directly indicated the formation of new chlorine- and lead-containing structural units. It was suggested that such units could be presented as Cl$^-$Pb^{2+}[BO$_{4/2}$]$^-$ complexes [2]. This composition of structural units could be explained by the fact that chlorine entered the glass network only in interstitial positions and acted as a non-bridging anion in the glass network [2,25]. Because the strength of an ionic bond is inversely proportional to the square of the ionic radius, substitution of O^{2-} with $r_{O^{2-}}$ = 1.40 Å by Cl$^-$ with r_{Cl^-} = 1.81 Å further weakened the bonds in the glass network. Thus, depolymerization and weakening of the glass network occurred with the formation of isolated ion-containing structural units.

Table 1. Interpretation of Raman bands in glasses.

Range of 80–250 cm^{-1}			Range of 250–1750 cm^{-1}		
Wavenumber, cm^{-1}	Interpretation	Reference	Wavenumber, cm^{-1}	Interpretation	Reference
80	Bending vibrations in structural units of PbO$_4$	[23,30]	310	Vibrations of [PbO$_4$]$^{2-}$ units	[23,31]
90	Stretching of predominantly ionic Pb–O bonds in polyhedral PbO$_n$ units	[23,24]	476	Pb–O bond vibrations	[32,33]
			570	Oscillations of borate tetrahedrons BØ$_4^-$	[23,25]
91	Vibrations associated with PbCl$_2$, presumably Cl$^-$	[21,22,27,34,35]	620	Deformation modes of BO$_3$ metaborate chains	[23,26]
			710	BO$_3$ deformation modes in ring metaborate groups	[23,26]
105	Cations Pb^{2+}	[25,28,36]	870	Oscillations BØ$_4^-$ in pentaborate groups	[26,37]
120	Pb–Cl stretching vibration	[22,29]	910	Oscillations BØ$_4^-$ in ortho- and pentaborate groups	[26,37]
135	Symmetric Pb–O stretching in the [PbO$_4$]$^{2-}$ pyramid configuration	[25,26,32,38]	980	Oscillations BØ$_4^-$ of diborate groups	[26,39,40]
150–180	Valence vibrations of Pb–Cl, bending vibrations of the Pb–Cl bond	[22,27,32,34,35,41]	1050	Oscillations BØ$_4^-$ of diborate groups	[23,39,40]
			1075	Oscillations BØ$_4^-$ of diborate groups	[23,39]
			1285	B–O$^-$ stretches in metaborate triangles (BØ$_2$O$^-$), mostly forming chain structures. A minor fraction: BØ$_3$ extensions	[23,25,33]

We observed several characteristic bands related to different groups in the glass structure in the 250–1750 cm^{-1} range (Figure 6). An increase in PbCl$_2$ content resulted in a decrease in band intensities.

The band around 310 cm^{-1} indicated the existence of PbO$_4$ polyhedral units in the network [23]. This was caused by the superposition of two bands centered at 280 and 330 cm^{-1}, which were attributed to vibrations of these units in PbO crystals with orthorhombic and tetragonal symmetry, respectively [18]. The decrease in the band intensity indicated the destruction of PbO$_4$ tetrahedra in the network with PbCl$_2$ content growth. The band at 476 cm^{-1} was also assigned to Pb–O bond vibrations and it underwent similar changes [33].

The 570 cm^{-1} band was associated with the borate tetrahedron (BØ$_4^-$) and it decreased with PbCl$_2$ content growth. The BO$_3$—PbO$_4$ bridge bond formed a fourfold coordination of boron with oxygen [23].

Bands at 620 and 710 cm^{-1} were assigned to the deformation modes of metaborate chains and rings, respectively [23,26,37]. They also lost their intensity with PbCl$_2$ content growth. However, the band at 710 cm^{-1} conserved its shape even for systems with 40 mol% of lead chloride, which indicated greater stability of the ring metaborate groups.

The 870 and 910 cm^{-1} bands decreased while the PbCl$_2$ content grew to 20 mol% and then remained constant with further PbCl$_2$ content growth. The 980, 1050, and 1075 cm^{-1} bands referred to the presence of BØ$_4^-$ in diborate groups and showed a similar evolution with PbCl$_2$ content growth.

The 1285 cm^{-1} band referred to B-O$^-$ extensions in metaborate triangles (BØ$_2$O$^-$), mostly forming chain structures, and to BØ$_3$ stretching. The band intensity strongly decreased with PbCl$_2$ content growth.

For the 50-50 glass composition, the intensity of the bands related to the BO$_4^-$ vibrations was noticeably lower than for the bands related to the BO$_3$ vibrations, primarily at 710 and 1285 cm^{-1}. We observed a reduction in the I_{710}/I_{1285} ratio with PbCl$_2$ content growth resulting from destruction of the Pb-O bond. The glass network was rearranged to a more ionic and depolymerized one, with the gradual replacement of BO$_3$ groups by BO$_4$ groups [25].

In the range up to 3500 cm^{-1} (Figure S4), only water oscillations were observed.

The FT-IR spectra showed several bands (Figures 7 and S5), interpreted in Table 2.

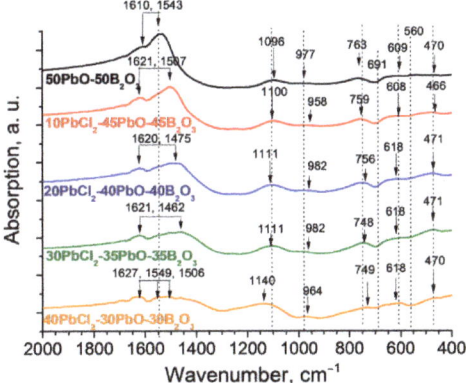

Figure 7. FT-IR spectra of synthesized glasses.

The 470 cm^{-1} band referred to the vibrations of Pb^{2+} cations and its intensity increased with PbCl$_2$ content growth. The 660 cm^{-1} band referred to vibrations of free BO$_4$ groups. It was practically absent in the 50-50 glass but it manifested itself and increased with PbCl$_2$ content growth.

The 691 cm^{-1} band was associated with B-O vibrations in the PbO—BO$_3$ bridge and the band intensity rapidly decreased with PbCl$_2$ content growth, which directly indicated the destruction of PbO$_4$ tetrahedra.

Peaks in the 800–1200 cm^{-1} range referred to vibrations of the B–O bond in units of BO$_4$. The bands were retained and even slightly increased with PbCl$_2$ content growth, which indicated the retention of BO$_4$ units in the grid.

Peaks in the 1300–1700 cm^{-1} range were attributed to bond fluctuations in the BO$_3$ groups. Their intensities decreased with PbCl$_2$ content growth, which indicated the destruction of the corresponding structural units.

We should note that the FT-IR data were in full agreement with the Raman spectroscopy data. PbCl$_2$ content growth resulted in the generation of new lead–chloride structural-chemical units, accompanied by a significant decrease in concentrations of BO$_3$ groups and PbO$_4$ tetrahedra.

Table 2. Interpretation of FT-IR bands in glasses.

Wavenumber, cm^{-1}	Corresponding Vibrational Mode	Reference
~466	Lead cation oscillations Pb^{2+}	[42,43]
~560	Oscillations of free BO$_4$ groups	[44]
~610	bending vibrations of the B–O–B bonds in the BO$_3$ group	[45]
~691	B-O oscillations associated with the PbO—BO$_3$ bridge in the lead borate network	[45,46]
~760	Bending vibrations of the O$_3$B–O–BO$_4$ bond	[5]
~800–1200	Tensile vibrations of the B–O bonds of the tetrahedral block BO$_4$ in ortho-, pyro-, and metaborate groups	[5,45,46]
970	Tensile vibrations of B–O bonds in BO$_4$ groups	[45,46]
~1050–1150	Tensile B-O oscillations in BO$_4$ units from tri-, tetra-, and pentaborate groups	[4,5]
1300–1700	Bond fluctuations in BO$_3$ groups	[4,5,45,46]
~1505, 1559, 1611	Asymmetric relaxations of B–O bond stretching of BO$_3$ trigonal blocks	[44]
~2350	Asymmetric modes of CO$_2$ stretching	[47]
~3445	Stretching vibrations of OH$^-$ groups	[4,44,47]

3.2. DSC Characterization and Physical Properties

The presented DSC data (Figures 7, S6–S10, Table 3) led us conclude that for the glasses of this system, the glass transition temperature (T_g) decreased with PbCl$_2$ content growth. For the boundaries of PbCl$_2$ composition, the difference was 80 °C. Such a significant change in the glass transition temperature (T_g) has been observed by other researchers and was attributed to structural changes in the glass network [16,48,49].

Table 3. Glass characteristic temperatures *.

Glass Composition, mol%	T_g, °C	T_{x1}, °C	T_{c1}, °C	T_{x2}, °C	T_{c2}, °C	T_{m1}, °C	T_{m2}, °C	$\Delta T = (T_{x1} - T_g)$, °C
50PbO–50B$_2$O$_3$	331	-	-	-	-	-	-	-
10PbCl$_2$—5PbO–45B$_2$O$_3$	327	405	427	485	540	522	601	78
20PbCl$_2$–40PbO–40B$_2$O$_3$	303	402	447	-	-	383	604	99
30PbCl$_2$–35PbO–35B$_2$O$_3$	279	408	436	-	-	390	470	129
40PbCl$_2$–30PbO–30B$_2$O$_3$	251	381	397	428	-	381	467	130

*—the determination error for all characteristic temperatures was ±1 °C.

The incorporation of halogen ions (Cl, Br, I) into the gaps of the borate network led to a violation of the order of the BO$_3$ groups. The overall disorder in halogen-doped glass occurred mainly due to the expansion of the boron–oxygen network [50,51]. Glasses with PbCl$_2$ concentrations < 20 mol% consisted mainly of BO$_3$ units, which had a relatively open structure. The large Cl$^-$ ions could be easily incorporated in the network only in interstitial positions. With the further addition of PbCl$_2$ (>20 mol%), the B–O network began to expand due to PbCl$_2$. The network retained its connectivity, but it contained large voids that allowed the placement of Cl$^-$ ions. The expansion of the B–O network, and hence the progressive weakening of the structure, manifested itself in a significant decrease in the glass transition temperature (T_g) [16]. This mechanism was confirmed by our structural studies.

At the same time, the DSC curves showed that the characteristics of crystallization in the studied glasses changed. The 50PbO–50B$_2$O$_3$ sample did not exhibit an exothermic peak corresponding to crystallization. However, we observed a broadened exothermic peak of crystallization with PbCl$_2$ content growth. A further increase in PbCl$_2$ content resulted in the splitting of the peak into two components. At 40 mol% PbCl$_2$, we observed only the crystallization peak with a lower T_c.

The lead borate oxychloride glasses demonstrated two endothermic peaks at continuous heating, marked as T_{m1} and T_{m2} (Figure 8). This indicated the stepwise nature of the glass softening. We observed the final melting temperature of the studied glasses at ≤600 °C, which made it possible to classify these glasses as low-melting glasses (Figures S6–S10).

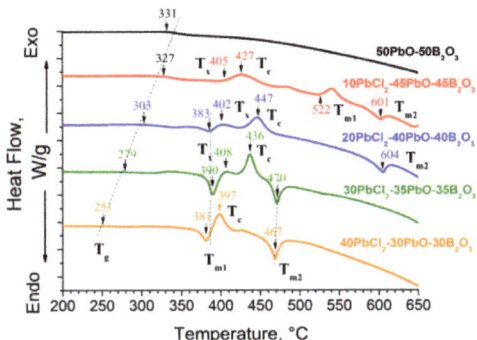

Figure 8. DSC curves of synthesized glasses.

One of the most important criteria for glasses is the thermal stability coefficient (ΔT), defined as the difference between the onset of crystallization (T_x) and glass transition (T_g) [52] temperatures (Table 3). For all glass compositions, ΔT was higher than 70 °C. This meant that the studied glasses were resistant to crystallization. An increase in ΔT indicated an increase in glass stability with $PbCl_2$ content growth [53–55].

According to the TGA data, volatilization of the studied glasses began in the temperature region of 540–630 °C, and $PbCl_2$ content growth intensified the volatilization (Figure 9). In contrast, the $50PbO-50B_2O_3$ glass did not volatilize up to 800 °C. Since the crystallization of the corresponding compositions occurred before active volatilization had begun, one could conclude that it was possible to create glass-crystalline materials based on glasses without changing the chemical composition.

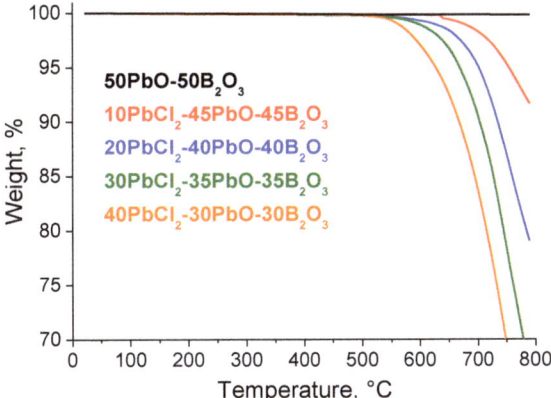

Figure 9. Results of thermogravimetric analysis of synthesized glasses.

Based on the data from measuring the density (ρ) of the glass samples, the molar volume V_m was calculated using formula (3):

$$V_m = M/\rho, \tag{3}$$

where M is the average molecular weight of the glass [9].

The glass density gradually decreased with increasing $PbCl_2$ content in glass (Figure 10). In similar oxychloride glass systems [5,16,49], the same dependencies were observed.

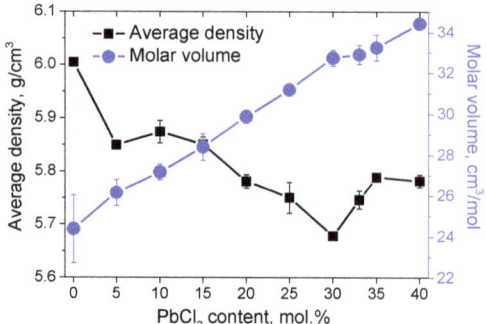

Figure 10. Dependence of density (black) and molar volume (blue) on the content of PbCl$_2$ in xPbCl$_2$–(50-0.5x)PbO–(50-0.5x)B$_2$O$_3$ glass.

An increase in PbCl$_2$ concentration caused the density decrease and molar volume growth that were associated with a change in the structure of the glass network [5]. The big Cl$^-$ ion occupies more space than the smaller O^{2-} and expands the glass network, thereby increasing the molar volume [9,12], which was consistent with the data on the glass structure.

Glass samples containing a larger amount of PbCl$_2$ had lower Vickers hardness (Figures 11 and S11). Similar results for the same glass systems were presented in [14,48]. The transition from triangular BO$_3$ groups to tetrahedral BO$_4$ groups reduced the hardness and increased the non-bridging oxygens in binary [56] and ternary [57] borate glass systems. The trend towards a decrease in Vickers hardness with increased PbCl$_2$ composition was associated with the formation of a larger number of depolymerizable boron structural units and a decrease in the rigidity of covalent bonds within them [58]. According to [20], the Vickers hardness of lead borate glasses mainly depends on the nature of the types of lead and boron structural units present in the glasses.

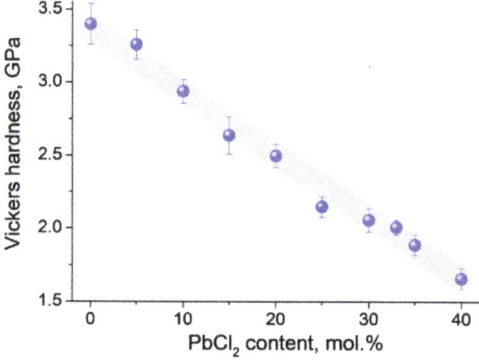

Figure 11. Dependence of Vickers hardness on the content of PbCl$_2$ in xPbCl$_2$–(50-0.5x)PbO–(50-0.5x)B$_2$O$_3$ glass. The line is guide for the eyes.

3.3. Optical Properties

The optical absorption spectra obtained in the UV, visible, and near-IR regions of the glasses under study are shown in Figure 12. The optical absorption coefficient, $\alpha(\lambda)$, was calculated from the absorption, $A(\lambda)$, based on the following relation [9]:

$$\alpha(\lambda) = 2.303 \frac{A(\lambda)}{d}, \tag{4}$$

where d is the thickness of the glass sample.

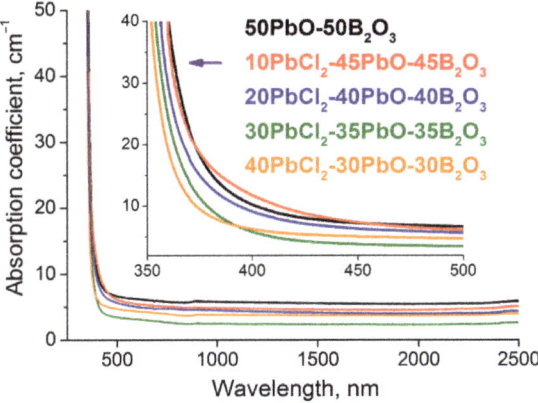

Figure 12. Optical absorption spectra of synthesized glasses. Enlarged fragment in the inset.

On the other hand, the optical absorption coefficient, $\alpha(\nu)$, for amorphous materials is given by the Tauc equation [13]:

$$\alpha(\nu) = const\left[\frac{(h\nu - E_g)^2}{h\nu}\right], \qquad (5)$$

where $h\nu$ is the photon energy and E_g is the optical energy gap. Thus, based on the Tauc plots (Figure 13), the optical gap (E_g) of the studied glasses was estimated.

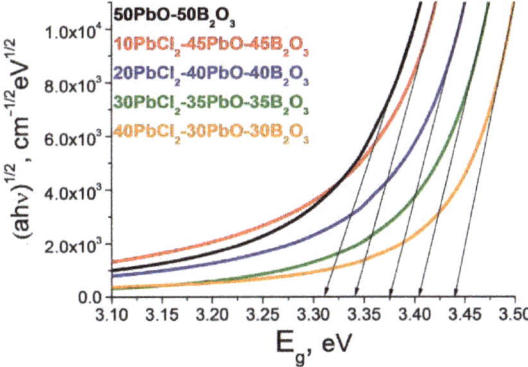

Figure 13. Tauc plots of optical absorption spectra of synthesized glasses.

The Tauc energy (optical band gap) increased linearly with an increase in the PbCl$_2$ content (Figure 14). This was the reason for the shift of the short-wavelength absorption edge of the glasses to the UV region (inset in Figure 12, Table 4).

The absorption edge in oxide glasses corresponds to the transition of an oxygen electron to the excited state [59]. The more weakly these oxygen electrons are bound, the more easily absorption occurs. In glasses, non-bridging oxygen is more negative than bridging oxygen. Increasing the proportion of ionized oxygen by converting it from bridging to non-bridging oxygen raises the valence band top, which leads to a decrease in the band gap (E_g) [60].

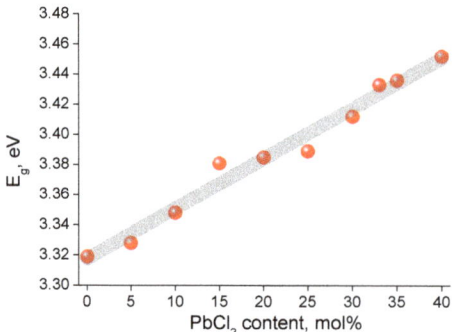

Figure 14. Dependence of the optical band gap on the PbCl$_2$ content in glasses. The line provides a guide for the eyes.

Table 4. Optical properties of glasses.

Glass Composition, mol%	Short-Wavelength Absorption Edge λ, nm	Tauc Energy E_g, eV *	Urbach Energy E_e, eV *	Average Absorption Coefficient, cm^{-1}	Long-Wavelength Absorption Edge λ, nm	Absorption of the OH$^-$ Group Band, cm^{-1}
50PbO–50B$_2$O$_3$	365	3.307	0.468	5.7	3611	15.5
5PbCl$_2$–47.5PbO–47.5B$_2$O$_3$	365	3.320	0.393	3.6	3626	22.0
10PbCl$_2$–45PbO–45B$_2$O$_3$	363	3.327	0.504	4.7	3664	22.4
15PbCl$_2$–42.5PbO–42.5B$_2$O$_3$	360	3.378	0.432	4.4	3690	20.3
20PbCl$_2$–40PbO–40B$_2$O$_3$	360	3.376	0.435	5.0	3739	12.4
25PbCl$_2$–37.5PbO–37.5B$_2$O$_3$	359	3.389	0.505	5.2	3788	11.9
30PbCl$_2$–35PbO–35B$_2$O$_3$	357	3.406	0.377	2.4	4349	14.6
33PbCl$_2$–33PbO–34B$_2$O$_3$	356	3.414	0.453	4.6	4370	10.9
35PbCl$_2$–32.5PbO–32.5B$_2$O$_3$	356	3.423	0.434	4.4	4462	9.5
40PbCl$_2$–30PbO–30B$_2$O$_3$	355	3.440	0.427	3.8	4710	9.4

*—the determination errors for E_g and E_e were ±0.005 eV.

In our case, the increase in the optical energy gap (E_g) with PbCl$_2$ content growth could be explained by a decrease in the content of non-bridging oxygen together with BO$_3$ groups. On the other hand, the expansion of the band gap can be explained in terms of the E_g values of the initial glass components: E_g^{PbO} = 2.8 eV; $E_g^{B_2O_3}$ = 5.4 eV; and $E_g^{PbCl_2}$ = 4.9 − 5.0 eV [61].

The absorption coefficient, $\alpha(\nu)$, in amorphous materials in the optical region near the absorption edge at a certain temperature obeys an empirical relation known as the Urbach rule [62]:

$$\alpha(\nu) = \alpha_0 exp(h\nu/E_e), \qquad (6)$$

where $h\nu$ is the photon energy, α_0 is a constant, and E_e is the energy, which is interpreted as the width of the localized state in the normal bandgap known as the Urbach energy. The Urbach energy values were obtained by finding the reciprocal of the tangent of the slope of the graph plotted in the $ln(\alpha(\nu)) - h\nu$ coordinates and are presented in Table 4.

The origin of exponential absorption is still a matter of debate, but it is generally believed that random potential fluctuations are associated with any lattice distortions, which can affect the Urbach energy by the generation of energy states within the band gap [13]. The nonlinear change in the Urbach energy depending on the PbCl$_2$ concentration may have been due to the uneven distribution of defects in the studied glasses. Similar results for the Urbach energy were presented in [5,9].

For the mid-IR spectrum range, we found that an increase in PbCl$_2$ concentration shifted the long-wavelength absorption edge to the IR region (Figure 15, Table 4) due to the replacement of atoms with a low atomic number (B, Z_B = 5) by atoms with a high atomic number (Pb, Z_{Pb} = 82). The glass matrix became heavier when we increased the total concentration of lead compounds (PbCl$_2$ and PbO). This resulted in the shift of the long-wavelength absorption edge in the IR region and the expansion of the transparency range of the glasses [63]. The addition of a low-energy phonon component (PbCl$_2$ is 230 cm^{-1} compared to borate glass ~1500 cm^{-1}) also increased the transparency range in the IR region.

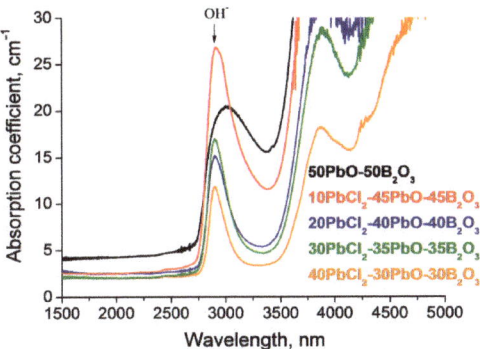

Figure 15. Optical absorption spectra of synthesized glasses in the range from 1500 to 5000 nm.

We observed a decrease in the intensity of the absorption band of hydroxyl groups with PbCl$_2$ content growth (Figure 15, Table 4). According to the FT-IR spectroscopy results of the starting reagents, B$_2$O$_3$ had the largest content of OH$^-$ groups (Figure S12). To reduce the content of OH$^-$ groups, it is possible to use dry boron oxide [64] and bubble the melt with dry oxygen [65].

The theoretical refractive index of the studied glasses, (n), was calculated using Equation (7) [38]:

$$n = \chi(PbCl_2) \cdot n(PbCl_2) + \chi(PbO) \cdot n(PbO) + \chi(B_2O_3) \cdot n(B_2O_3), \tag{7}$$

where $\chi(PbCl_2)$, $\chi(PbO)$, and $\chi(B_2O_3)$ are the molar fractions of the corresponding glass components; $n(PbCl_2)$, $n(PbO)$, and $n(B_2O_3)$ are the refractive indices of pure lead chloride, lead oxide, and boron oxide, respectively.

The refractive index was determined experimentally at three wavelengths: 633, 969, and 1539 nm (Table 5). The dispersions of the refractive index depending on the wavelength are presented in Figure S13.

Table 5. The values of the calculated and experimental refractive indices (n) of glasses in the xPbCl$_2$–(50-0.5x)PbO–(50-0.5x) system.

Glass Composition, mol%	Refractive Indices (n)			
	Estimated, ±0.01	Experimental, ±0.001		
		633 nm	969 nm	1539 nm
50PbO–50B$_2$O$_3$	2.04	1.914	1.895	1.890
5PbCl$_2$–47.5PbO–47.5B$_2$O$_3$	2.05	1.969	1.930	1.903
10PbCl$_2$–45PbO–45B$_2$O$_3$	2.06	1.989	1.949	1.932
15PbCl$_2$–42.5PbO–42.5B$_2$O$_3$	2.06	1.991	1.963	1.940
20PbCl$_2$–40PbO–40B$_2$O$_3$	2.07	2.009	1.969	1.954
25PbCl$_2$–37.5PbO–37.5B$_2$O$_3$	2,08	2.017	1.978	1.960
30PbCl$_2$–35PbO–35B$_2$O$_3$	2.09	2.040	2.000	1.981
33PbCl$_2$–33PbO–34B$_2$O$_3$	2.09	2.051	2.010	1.991
35PbCl$_2$–32.5PbO–32.5B$_2$O$_3$	2.10	2.058	2.016	1.997
40PbCl$_2$–30PbO–30B$_2$O$_3$	2.10	2.076	2.032	2.013

The values of the calculated refractive index were slightly overestimated relative to the experimental ones. We assumed that this was due to some difference in the nominal and actual compositions of the glasses, primarily due to the incorporation of Al_2O_3 (n = 1.70–0.76) into the glasses, as determined by XPS [66].

4. Conclusions

For the first time we investigated the properties of glasses in the $xPbCl_2$–$(50-0.5x)PbO$–$(50-0.5x)B_2O_3$ system with a wide range of $PbCl_2$ concentrations from 0 to 40 mol%. This allowed us to fill gaps in the fundamental data for the system under study, which has applications for the production of optical fibers with extended transmission ranges for use in medical lasers and telecommunication systems.

The introduction of $PbCl_2$ modified the structural network and led to drastic changes in the glassy network, resulting in a significant decrease in the numbers of BO_3 groups and PbO_4 tetrahedra, T_g reduction by 80 °C with a simultaneous increase in resistance to crystallization by 52 °C, glass molar volume growth of 30%, Vickers hardness reduction by 2 times, and refractive index growth from 1.9 to 2.1.

The decrease in bridge oxygen content with $PbCl_2$ content growth resulted in the expansion of transparency in the UV region by 10 nm. The increase in heavy metal content in the glass matrix resulted in a decrease in the phonon energy, which resulted in the expansion of the IR range to 4700 nm.

Supplementary Materials: The following supporting information can be downloaded at: https://www.mdpi.com/article/10.3390/ceramics6030083/s1, Figure S1. Comparison of XRD patterns of glasses in the $xPbCl_2$–$(50-0.5x)PbO$–$(50-0.5x)B_2O_3$ system with crystalline $PbCl_2$; Table S1. The composition of glasses according to the XFS data; Figure S2. Decomposition of the $50PbO$–$50B_2O_3$ glass Raman spectrum into Gaussian; Figure S3. Decomposition of the $40PbCl_2$–$30PbO$–$30B_2O_3$ glass Raman spectrum into Gaussian; Figure S4. Raman spectra of $xPbCl_2$–$(50-0.5x)PbO$–$(50-0.5x)B_2O_3$ glasses (1750–3500 cm^{-1} range); Figure S5. FT-IR spectra of synthesized glasses; Figure S6. DSC curve for glass composition $50PbO$–$50B_2O_3$ and characteristic temperatures; Figure S7. DSC curve for glass composition $10PbCl_2$–$45PbO$–$45B_2O_3$ and characteristic temperatures; Figure S8. DSC curve for glass composition $20PbCl_2$–$40PbO$–$40B_2O_3$ and characteristic temperatures; Figure S9. DSC curve for glass composition $30PbCl_2$–$35PbO$–$35B_2O_3$ and characteristic temperatures; Figure S10. DSC curve for glass composition $40PbCl_2$–$30PbO$–$30B_2O_3$ and characteristic temperatures; Figure S11. Dependence of Vickers hardness on the applied load in $xPbCl_2$–$(50-0.5x)PbO$–$(50-0.5x)B_2O_3$ glasses; Figure S12. FT-IR spectra of the initial reagents, with the indicated band of the OH^- group; Figure S13. The dispersion of the refractive index of $xPbCl_2$–$(50-0.5x)PbO$–$(50-0.5x)B_2O_3$ glasses.

Author Contributions: Conceptualization, D.B. and O.P.; methodology, D.B., K.R. and O.P.; software, A.B.; validation, O.P. and I.A.; formal analysis, M.U.; investigation, D.B., A.B., K.R., I.K., M.U. and A.P.; resources, K.R, A.P., M.U. and I.A.; data curation, D.B. and O.P.; writing—original draft preparation, D.B., A.B., O.P. and I.A.; writing—review and editing, O.P. and I.A.; visualization, D.B., I.K., A.P. and M.U.; supervision, I.A.; project administration, O.P.; funding acquisition, K.R. All authors have read and agreed to the published version of the manuscript.

Funding: The research was financially supported by the Ministry of Science and High Education of Russia by the project FSSM-2020-0005.

Institutional Review Board Statement: Not applicable.

Informed Consent Statement: Not applicable.

Data Availability Statement: Not applicable.

Acknowledgments: Analytical studies were carried out using the scientific equipment of the Central Collective Use Center of the National Research Center «Kurchatov Institute»—IREA.

Conflicts of Interest: The authors declare no conflict of interest.

References

1. Bengisu, M. Borate glasses for scientific and industrial applications: A review. *J. Mater. Sci.* **2016**, *51*, 2199–2242. [CrossRef]
2. Sokolov, I.A.; Murin, I.V.; Mel'Nikova, N.A.; Pronkin, A.A. A Study of Ionic Conductivity of Glasses in the $PbCl_2$–PbO · B_2O_3 and $PbCl_2$–2PbO · B_2O_3 Systems. *Glas. Phys. Chem.* **2003**, *29*, 291–299. [CrossRef]
3. Singh, G.P.; Singh, J.; Kaur, P.; Singh, T.; Kaur, R.; Singh, D.P. The Role of Lead Oxide in PbO-B_2O_3 Glasses for Solid State Ionic Devices. *Mater. Phys. Mech.* **2021**, *47*, 951–961. [CrossRef]
4. Pisarska, J. Novel oxychloroborate glasses containing neodymium ions: Synthesis, structure and luminescent properties. *J. Mol. Struct.* **2008**, *887*, 201–204. [CrossRef]
5. Sekhar, K.C.; Hameed, A.; Sathe, V.G.; Chary, M.N.; Shareefuddin, M. Physical, optical and structural studies of copper-doped lead oxychloro borate glasses. *Bull. Mater. Sci.* **2018**, *41*, 79. [CrossRef]
6. Saddeek, Y.B. Structural and acoustical studies of lead sodium borate glasses. *J. Alloys Compd.* **2009**, *467*, 14–21. [CrossRef]
7. Yousef, E.S.S. Thermal and optical properties of zinc halotellurite glasses. *J. Mater. Sci.* **2007**, *42*, 4502–4507. [CrossRef]
8. Brown, E.; Hömmerich, U.; Bluiett, A.G.; Trivedi, S.B.; Zavada, J.M. Synthesis and spectroscopic properties of neodymium doped lead chloride. *J. Appl. Phys.* **2007**, *101*, 113103. [CrossRef]
9. Sekhar, K.C.; Hameed, A.; Ramadevudu, G.; Chary, M.N. Shareefuddin Physical and spectroscopic studies on manganese ions in lead halo borate glasses. *Mod. Phys. Lett. B* **2017**, *31*, 1750180. [CrossRef]
10. Pisarski, W.A.; Pisarska, J.; Lisiecki, R.; Grobelny, Ł.; Dominiak-Dzik, G.; Ryba-Romanowski, W. Erbium-doped oxide and oxyhalide lead borate glasses for near-infrared broadband optical amplifiers. *Chem. Phys. Lett.* **2009**, *472*, 217–219. [CrossRef]
11. Pisarska, J. Borate glasses with PbO and $PbCl_2$ containing Dy^{3+} ions. *Opt. Appl.* **2010**, *40*, 367–374.
12. El-Damrawi, G. Influence of $PbCl_2$ on physical properties of lead chloroborate glasses. *J. Non-Cryst. Solids* **1994**, *176*, 91–97. [CrossRef]
13. Sebastian, S.; Khadar, M.A. Optical properties of $60B_2O_3$-(40-x)PbO-$xMCl_2$ and $50B_2O_3$-(50-x) PbO-$xMCl_2$ (M = Pb, Cd) glasses. *Bull. Mater. Sci.* **2004**, *27*, 207–212. [CrossRef]
14. Sebastian, S.; Khadar, M.A. Microhardness indentation size effect studies in $60B_2O_3$-(40-x)PbO-$xMCl_2$ and $50B_2O_3$(50-x)PbO-$xMCl_2$ (M = Pb, Cd) glasses. *J. Mater. Sci.* **2005**, *40*, 1655–1659. [CrossRef]
15. Menezes, D.; Bannwart, E.; De Souza, J.; Rojas, S. Thermoluminescence emission on lead oxychloroborate glasses under UV exposure. *Luminescence* **2019**, *34*, 918–923. [CrossRef] [PubMed]
16. Silim, H.A.; El-Damrawi, G.; Moustafa, Y.M.; Hassan, A.K. Electrical and elastic properties of binary lead chloroborate glasses. *J. Phys. Condens. Matter* **1994**, *6*, 6189–6196. [CrossRef]
17. Nunes, C.; Mahendrasingam, A.; Suryanarayanan, R. Quantification of Crystallinity in Substantially Amorphous Materials by Synchrotron X-ray Powder Diffractometry. *Pharm. Res.* **2005**, *22*, 1942–1953. [CrossRef] [PubMed]
18. Bates, S.; Zografi, G.; Engers, D.; Morris, K.; Crowley, K.; Newman, A. Analysis of Amorphous and Nanocrystalline Solids from Their X-ray Diffraction Patterns. *Pharm. Res.* **2006**, *23*, 2333–2349. [CrossRef]
19. Cattaneo, A.S.; Lima, R.P.; Tambelli, C.E.; Magon, C.J.; Mastelaro, V.R.; Garcia, A.; de Souza, J.E.; de Camargo, A.S.S.; de Araujo, C.C.; Schneider, J.F.; et al. Structural Role of Fluoride in the Ion-Conducting Glass System B_2O_3–PbO–LiF Studied by Single- and Double-Resonance NMR. *J. Phys. Chem. C* **2008**, *112*, 10462–10471. [CrossRef]
20. Singh, G.P.; Kaur, P.; Kaur, S.; Singh, D. Role of WO_3 in structural and optical properties of WO_3–Al_2O_3–PbO–B_2O_3 glasses. *Phys. B Condens. Matter* **2011**, *406*, 4652–4656. [CrossRef]
21. Tsai, P.; Cooney, R.P. Raman spectra of polynuclear hydroxo-compounds of lead(II) chloride. *J. Chem. Soc. Dalton Trans.* **1976**, 1631–1634. [CrossRef]
22. Melo, F.E.A.; Filho, J.M.; Moreira, J.E.; Lemos, V.; Cerdeira, F. Anharmonic effects in Raman-active modes of $PbCl_2$. *J. Raman Spectrosc.* **1984**, *15*, 128–131. [CrossRef]
23. Chatzipanagis, K.I.; Tagiara, N.S.; Kamitsos, E.I.; Barrow, N.; Slagle, I.; Wilson, R.; Greiner, T.; Jesuit, M.; Leonard, N.; Phillips, A.; et al. Structure of lead borate glasses by Raman, ^{11}B MAS, and ^{207}Pb NMR spectroscopies. *J. Non-Cryst. Solids* **2022**, *589*, 121660. [CrossRef]
24. Takaishi, T.; Jin, J.; Uchino, T.; Yoko, T. Structural Study of PbO-B_2O_3 Glasses by X-ray Diffraction and ^{11}B MAS NMR Techniques. *J. Am. Ceram. Soc.* **2004**, *83*, 2543–2548. [CrossRef]
25. Pan, Z.; Henderson, D.O.; Morgan, S.H. A Raman investigation of lead haloborate glasses. *J. Chem. Phys.* **1994**, *101*, 1767–1774. [CrossRef]
26. Filho, A.S.; Filho, J.M.; Melo, F.; Custódio, M.; Lebullenger, R.; Hernandes, A. Optical properties of Sm^{3+} doped lead fluoroborate glasses. *J. Phys. Chem. Solids* **2000**, *61*, 1535–1542. [CrossRef]
27. Ozin, G.A. The single crystal Raman spectrum of orthorhombic $PbCl_2$. *Can. J. Chem.* **1970**, *48*, 2931–2933. [CrossRef]
28. Mendes-Filho, J.; Melo, F.E.; Moreira, J.E. Raman spectra of lead chloride single crystals. *J. Raman Spectrosc.* **1979**, *8*, 199–202. [CrossRef]
29. Sun, H.-T.; Wen, L.; Duan, Z.-C.; Hu, L.-L.; Zhang, J.-J.; Jiang, Z.-H. Intense frequency upconversion fluorescence emission of Er^{3+}/Yb^{3+}-codoped oxychloride germanate glass. *J. Alloys Compd.* **2006**, *414*, 142–145. [CrossRef]
30. Witke, K.; Hübert, T.; Reich, P.; Splett, C. Quantitative Raman investigations of the structure of glasses in the system B_2O_3-PbO. *Phys. Chem. Glas.* **1994**, *35*, 28–33.

31. Adams, D.M.; Stevens, D.C. Single-crystal vibrational spectra of tetragonal and orthorhombic lead monoxide. *J. Chem. Soc. Dalton Trans.* **1977**, 1096–1103. [CrossRef]
32. Frost, R.L.; Williams, P.A. Raman spectroscopy of some basic chloride containing minerals of lead and copper. *Spectrochim. Acta Part A: Mol. Biomol. Spectrosc.* **2004**, *60*, 2071–2077. [CrossRef] [PubMed]
33. Sekhar, K.C.; Rao, G.V.; Chary, B.S.; Phani, A.V.L.; Sathe, V.G.; Narsimlu, N. Shareefuddin Raman study on lead halo borate glasses: Effect of PbO as modifier. In Proceedings of the International Conference on Advanced Materials: ICAM 2019, Kerala, India, 12–14 June 2019; p. 020144. [CrossRef]
34. Zakir'yanov, D.O.; Chernyshev, V.; Zakir'yanova, I.D. Phonon spectrum of lead oxychloride $Pb_3O_2Cl_2$: Ab initio calculation and experiment. *Phys. Solid State* **2016**, *58*, 325–332. [CrossRef]
35. Maroni, V.A.; Cunningham, P.T. Laser-Raman spectra of gaseous $BiBr_3$ and $PbCl_2$. *Appl. Spectrosc.* **1973**, *27*, 428–430. [CrossRef]
36. Faulques, E.; Zubkowski, J.D.; Perry, D.L. Infrared and Raman Spectra of bis-Thiourea Lead(II) Chloride. *Spectrosc. Lett.* **1996**, *29*, 1275–1284. [CrossRef]
37. Li, Z.; Sun, Y.; Liu, L.; Zhang, Z. Modification of the Structure of Ti-Bearing Mold Flux by the Simultaneous Addition of B_2O_3 and Na_2O. *Met. Mater. Trans. E* **2016**, *3*, 28–36. [CrossRef]
38. Kurushkin, M.; Semencha, A.; Blinov, L.N.; Mikhail, M. Lead-containing oxyhalide glass. *Glas. Phys. Chem.* **2014**, *40*, 421–427. [CrossRef]
39. Górny, A.; Kuwik, M.; Pisarski, W.A.; Pisarska, J. Lead Borate Glasses and Glass-Ceramics Singly Doped with Dy^{3+} for White LEDs. *Materials* **2020**, *13*, 5022. [CrossRef]
40. Meera, B.; Sood, A.; Chandrabhas, N.; Ramakrishna, J. Raman study of lead borate glasses. *J. Non-Cryst. Solids* **1990**, *126*, 224–230. [CrossRef]
41. Coleman, F.; Feng, G.; Murphy, R.W.; Nockemann, P.; Seddon, K.R.; Swadźba-Kwaśny, M. Lead(ii) chloride ionic liquids and organic/inorganic hybrid materials—A study of chloroplumbate(ii) speciation. *Dalton Trans.* **2013**, *42*, 5025–5035. [CrossRef]
42. Chukanov, N.V.; Chervonnyi, A.D. *Infrared Spectroscopy of Minerals and Related Compounds*; Springer Mineralogy; Springer International Publishing: Cham, Switzerland, 2016. [CrossRef]
43. Muthuselvi, C.; Anbuselvi, T.; Pandiarajan, S. Synthesis and Characterization of Lead (II) Chloride and Lead (II) Carbonate Nano. *Part. Rev. Res.* **2016**, *5*, 1–13.
44. Abdelghany, A.M. Combined DFT, Deconvolution Analysis for Structural Investigation of Copper–doped Lead Borate Glasses. *Open Spectrosc. J.* **2012**, *6*, 9–14. [CrossRef]
45. Abdullah, M.; Shafieza, S.; Kasim, A.; Hashim, A. The Effect of PbO on the Physical and Structural Properties of Borate Glass System. *Mater. Sci. Forum* **2016**, *846*, 177–182. [CrossRef]
46. Pisarski, W.A.; Pisarska, J.; Ryba-Romanowski, W. Structural role of rare earth ions in lead borate glasses evidenced by infrared spectroscopy: $BO_3 \leftrightarrow BO_4$ conversion. *J. Mol. Struct.* **2005**, *744–747*, 515–520. [CrossRef]
47. Ma, Y.; Yan, W.; Sun, Q.; Liu, X. Raman and infrared spectroscopic quantification of the carbonate concentration in K_2CO_3 aqueous solutions with water as an internal standard. *Geosci. Front.* **2021**, *12*, 1018–1030. [CrossRef]
48. Rao, K.J.; Rao, B.G.; Elliott, S.R. Glass formation in the system $PbO-PbCl_2$. *J. Mater. Sci.* **1985**, *20*, 1678–1682. [CrossRef]
49. El-Damrawi, G. $PbCl_2$ conducting glasses with mixed glass formers. *J. Phys. Condens. Matter* **1995**, *7*, 1557–1563. [CrossRef]
50. Gressler, C.A.; Shelby, J.E. Lead fluoroborate glasses. *J. Appl. Phys.* **1988**, *64*, 4450–4453. [CrossRef]
51. Coon, J.; Horton, M.; Shelby, J.E. Stability and crystallization of lead halosilicate glasses. *J. Non-Cryst. Solids* **1988**, *102*, 143–147. [CrossRef]
52. Pisarski, W.; Pisarska, J.; Goryczka, T.; Dominiak-Dzik, G.; Ryba-Romanowski, W. Influence of thermal treatment on spectroscopic properties of Er^{3+} ions in multicomponent InF_3-based glasses. *J. Alloys Compd.* **2005**, *398*, 272–275. [CrossRef]
53. Gressler, C.A.; Shelby, J.E. Properties and structure of $PbO-PbF_2-B_2O_3$ glasses. *J. Appl. Phys.* **1989**, *66*, 1127–1131. [CrossRef]
54. Cheng, Y.; Xiao, H.; Guo, W. Structure and crystallization kinetics of $PbO-B_2O_3$ glasses. *Ceram. Int.* **2007**, *33*, 1341–1347. [CrossRef]
55. Petrova, O.; Velichkina, D.; Zykova, M.; Khomyakov, A.; Uslamina, M.; Nischev, K.; Pynenkov, A.; Avetisov, R.; Avetissov, I.C. Nd/La, Nd/Lu-co-doped transparent lead fluoroborate glass-ceramics. *J. Non-Cryst. Solids* **2020**, *531*, 119858. [CrossRef]
56. Greenblatt, S.; Bray, P.J. Structural Aspects of B-O Network in Glassy Borates. *Glas. Phys. Chem.* **1967**, *8*, 312.
57. Konijnendijk, W.; Stevels, J. The structure of borate glasses studied by Raman scattering. *J. Non-Cryst. Solids* **1975**, *18*, 307–331. [CrossRef]
58. Abdullah, M.; Mohd, W.; Husin, H.W.; Kassim, A.; Yahya, N. The physical properties of lead borate ($PbO-B_2O_3$) glass. *KONAKA Konf. Akad.* **2015**, *2015*, 380–386.
59. Hogarth, C.A.; Assadzadeh-Kashani, E. Some studies of the optical properties of tungsten-calcium-tellurite glasses. *J. Mater. Sci.* **1983**, *18*, 1255–1263. [CrossRef]
60. Duffy, J.A. Ultraviolet transparency of glass: A chemical approach in terms of band theory, polarisability and electronegativity. *Glas. Phys. Chem.* **2001**, *42*, 151–157.
61. Butenkov, D.A.; Runina, K.I.; Petrova, O.B. Synthesis and Properties of Nd-Doped Chlorofluorosilicate Lead Glasses. *Glas. Ceram.* **2021**, *78*, 135–139. [CrossRef]
62. Urbach, F. The Long-Wavelength Edge of Photographic Sensitivity and of the Electronic Absorption of Solids. *Phys. Rev.* **1953**, *92*, 1324. [CrossRef]

63. Lezal, D.; Pedlikova, J.; Kostka, P.; Bludska, J.; Poulain, M.; Zavadil, J. Heavy metal oxide glasses: Preparation and physical properties. *J. Non-Cryst. Solids* **2001**, *284*, 288–295. [CrossRef]
64. Avetissov, I.C.; Petrova, O.B.; Anurova, M.O.; Khomyakov, A.V.; Akkuzina, A.A.; Taidakov, I.V. Mechanical and optical properties of hybrid materials based on inorganic glass matrix and organic metal complex phosphors. *J. Phys. Conf. Ser.* **2018**, *1045*, 012006. [CrossRef]
65. Richards, B.D.O.; Teddy-Fernandez, T.; Jose, G.; Binks, D.; Jha, A. Mid-IR (3–4 µm) fluorescence and ASE studies in Dy^{3+} doped tellurite and germanate glasses and a fs laser inscribed waveguide. *Laser Phys. Lett.* **2013**, *10*, 085802. [CrossRef]
66. French, R.H.; Müllejans, H.; Jones, D.J. Optical Properties of Aluminum Oxide: Determined from Vacuum Ultraviolet and Electron Energy-Loss Spectroscopies. *J. Am. Ceram. Soc.* **2005**, *81*, 2549–2557. [CrossRef]

Disclaimer/Publisher's Note: The statements, opinions and data contained in all publications are solely those of the individual author(s) and contributor(s) and not of MDPI and/or the editor(s). MDPI and/or the editor(s) disclaim responsibility for any injury to people or property resulting from any ideas, methods, instructions or products referred to in the content.

Article

Water-Glass-Assisted Foaming in Foamed Glass Production

Sonja Smiljanić *, Uroš Hribar, Matjaž Spreitzer and Jakob König *

Advanced Materials Department, Jožef Stefan Institute, Jamova cesta 39, 1000 Ljubljana, Slovenia; uros.hribar@ijs.si (U.H.); matjaz.spreitzer@ijs.si (M.S.)
* Correspondence: sonja.smiljanic@ijs.si (S.S.); jakob.konig@ijs.si (J.K.)

Abstract: The energy efficiency of buildings can be greatly improved by decreasing the energy embodied in installed materials. In this contribution, we investigated the possibility of foaming waste bottle glass in the air atmosphere with the addition of water glass, which would reduce the energy used in the production of foamed glass boards. The results show that with the increased addition of water glass, the crystallinity and the thermal conductivity decrease, however, the remaining crystal content prevents the formation of closed-porous foams. The added water glass only partly protects the carbon from premature oxidation, and the foaming mechanism in the air is different than in the argon atmosphere. The lowest obtained foam density in the air atmosphere is 123 kg m^{-3}, while the lowest thermal conductivity is 53 mW m^{-1} K^{-1}, with an open porosity of 50% for the sample obtained in the air, containing 12 wt% of water glass, 2 wt% of B_2O_3, 2 wt% AlPO$_4$ and 2 wt% K_3PO_4.

Keywords: foam glass; waste glass; thermal conductivity

1. Introduction

One of the main focuses of energy savings in the European Union (EU) is energy use related to buildings, making it crucial to improve construction materials by lowering the embodied energy and further improving buildings' energy efficiency. These actions include the development of greener thermal insulation materials, which are one of the key materials used in energy-efficient buildings. Foamed glass is considered a sustainable insulation material, as it can be made from waste glass and is stable on a long timescale. However, producing high-quality foamed glass with superior insulation properties is an energy-intensive process that could be improved by eliminating the step of remelting waste glass (remelting represents approx. 20% of energy use) [1]. In comparison to conventional thermal insulation materials used in the building sector, i.e., mineral wool and organic foams (EPS, XPS), the best foamed glass reaches similar thermal conductivity values (36 mW m^{-1} K^{-1} vs. 30–35 mW m^{-1} K^{-1}), while having much better mechanical properties and long-term stability [2,3].

To avoid the remelting step, foamed glass is often produced through direct foaming of a mixture of finely milled waste glass and foaming agents, such as carbon or carbonate [1–6]. Carbon-based foaming agents react with chemically bonded oxygen (present as polyvalent ions in higher oxidation states) in the glass and release gases. To reduce the oxidation dependency on the glass composition, foaming mixtures can be supplemented with oxidizing agents, such as Fe$_2$O$_3$, manganese oxide in various oxidation states, and sulfates [7–9]. During the foaming process, the metal oxides incorporate, fully or partly, into the glass structure, which can trigger undesired crystallization [9]. Since the foaming temperatures are low, typically 750–850 °C, the glass is not completely homogenized, which results in local fluctuations in glass composition and glass instabilities. Glasses with lower glass stability are more susceptible to crystallization, which can hinder expansion and result in an open-porous foamed glass. Moreover, the finely milled glass is itself prone to crystallization [10].

In our previous research [10], we have shown that container waste glass exhibits low glass stability in a mixture with carbon and manganese oxide (Mn$_3$O$_4$) under foaming

Citation: Smiljanić, S.; Hribar, U.; Spreitzer, M.; König, J. Water-Glass-Assisted Foaming in Foamed Glass Production. *Ceramics* 2023, 6, 1646–1654. https://doi.org/10.3390/ceramics6030101

Academic Editors: Georgiy Shakhgildyan and Michael I. Ojovan

Received: 31 May 2023
Revised: 13 July 2023
Accepted: 31 July 2023
Published: 2 August 2023

Copyright: © 2023 by the authors. Licensee MDPI, Basel, Switzerland. This article is an open access article distributed under the terms and conditions of the Creative Commons Attribution (CC BY) license (https://creativecommons.org/licenses/by/4.0/).

conditions. Glass with a common soda-lime-silica (SLS) composition manifested complex crystallization with the formation of several crystalline phases. It was evidenced that when re-melted, the glass stability against crystallization improved, and fewer crystalline phases formed. However, the improvement was minor. Thereafter, we showed that crystallization can be more effectively suppressed by the addition of selected additives (borax, B_2O_3, Al_2O_3). Furthermore, we introduced phosphates in the foamed glass mixture, which improved the homogeneity of the foams and increased the content of closed pores [11]. The addition of the phosphates also decreased the densities of the foams below 150 kg m^{-3}, and the thermal conductivity of these foams was in the range of 57–66 mW m^{-1} K^{-1}. These samples were prepared under an oxygen-free atmosphere, which is used in the industry but is related to a higher energy consumption due to under-stoichiometric gas burning [5].

This study aims to investigate the influence of water glass (WG) addition on the foaming process and the possibility of transferring the foaming process from an oxygen-free atmosphere to an air atmosphere. Water glass is a known additive used in industrial processes. It was shown that WG could be used as a single foaming agent or in combination with carbonaceous foaming agents [12–14]. The proposed mechanism of the foaming with WG is due to the decomposition of carbonates, which are formed when wet water-glass-coated glass powder is in contact with the air atmosphere [15]. This study focuses on the foaming process of container glass waste using a foaming/oxidizing agent couple, with the addition of various crystallization inhibitors and water glass, the latter possibly enabling the foaming process in the air atmosphere. The effects of the different additives and foaming atmosphere on the properties of the foamed samples, such as density, porosity, crystallinity, and thermal conductivity, were investigated and compared. The underlying reactions and their influence on the properties of the foams are discussed.

2. Experimental Section

Waste glass with a typical SLS composition [10] was mixed with foaming/oxidizing agent couple 0.5 wt% carbon black (CB, acetylene black, Alfa Aesar, Kandel, Germany) and 6.356 wt% Mn_3O_4 and with different amounts of water glass (12 and 24 wt%), Table 1. Mn_3O_4 was prepared from MnO_2 (99%, Bie & Berntsen, Rødovre, Denmark) by thermal treatment at 1250 °C for 4 h. The mixture was homogenized in yttria-stabilized zirconia (YSZ) jar with 10 mm YSZ balls at 250 rpm in a planetary ball mill for 35 min. As crystallization inhibitors, we used fluxing agents: B_2O_3, and phosphates: K_3PO_4 and $AlPO_4$, which were also previously used as supporting materials for foaming waste glass [9,16]. WG was then mixed with the homogenized mixture in an agate mortar. The mixture with WG was finally dried at 200 °C for 1 h.

Table 1. Composition of the mixtures of the glass, 0.5 wt% CB, 6.356 wt% Mn_3O_4, 12 wt% WG or 24 wt% WG, foamed in air and Ar atmosphere.

Sample	Additives (wt%)		
	B_2O_3	$AlPO_4$	K_3PO_4
12WG-2Al-2K	/	2	2
12WG-2B-2Al	2	2	/
12WG-2B-2A-2K	2	2	2
24WG-2Al-2K	/	2	2
24WG-2B-2Al	2	2	/
24WG-2B-2Al-2K	2	2	2

Small samples were prepared from these mixtures by uniaxial pressing (ϕ 12) of ~1 g of the batch. The samples were heat-treated at 865 °C for 10 min with a heating rate of 10 K min^{-1} in an Ar and air atmosphere in a laboratory electrical tube furnace (Protherm ASP11/150/500, Ankara, Turkey).

The thermal behavior of foaming mixtures during heat treatment was investigated using a thermo-gravimetric analyzer (TGA) coupled with a mass spectrometer (MS). Specifically, the instrument used was the NETZSCH STA 449 C/6/G Jupiter TGA coupled with an Aëoloss QMS 403 mass spectrometer. In the study, approximately 10 mg of a powder mixture was analyzed. The samples were subjected to heating at a rate of 10 K min^{-1} and reached a maximum temperature of 800 °C. The analysis was carried out under two different gas atmospheres: synthetic air and Ar (flow 50 mL min^{-1}).

Powder X-ray diffraction (XRD; Malvern PANalytical Empyrean diffractometer, Malvern, United Kingdom) using a Cu-Kα radiation source was employed to identify the crystalline phases present in the foamed samples. The XRD data were collected within the 2θ range of 10–70°, with a step size of 0.0263° and a time per step of 500 s. The obtained diffraction patterns were compared with the patterns in the Joint Committee on Powder Diffraction Standards (JCPDS) database using Highscore Plus software for phase identification.

The apparent density (ρ_{app}) of the resulting foams was determined using the Archimedes principle in water for small samples, with a measurement error of ±1%. The pycnometric volume of the small samples was determined by the Archimedes method in absolute ethanol, with a measurement error of ±2%. To eliminate air from the open pores, the small samples were submerged in boiling ethanol under reduced pressure. The submerged samples' volume was then measured in absolute ethanol using the Archimedes principle (V_{pyc}).

The porosity of the foamed samples was determined based on the apparent density, pycnometer density, and powder density (measured as 2.50 g cm^{-3}). The following equations were used to calculate the total porosity φ (measurement error ±1%), open porosity (OP), and closed porosity (CP) (measurement error ±2%):

$$\varphi = 1 - \frac{\rho_{app}}{\rho_{pow}} \quad (1)$$

$$CP = \frac{\rho_{app}}{\rho_{py}} - \frac{\rho_{app}}{\rho_{pow}} \quad (2)$$

$$OP = \varphi - CP \quad (3)$$

where the volume (V_{foam}) of foam is:

$$V_{foam} = V_{CP} + V_{OP} + V_{glass}.$$

For the large samples, the apparent density was calculated using the sample's dimensions (geometric volume from the sample's dimensions) and weight. The total porosity of the samples was calculated based on the apparent density using Equation (1). The closed porosity was determined using an Ultrapyc 5000 Foam instrument (Anton Paar GmbH, Graz, Austria).

The thermal conductivity of the large foam samples, which were cut into dimensions of 8 cm × 8 cm × 2 cm, was measured using a heat-flow meter (HFM 446 Lambda Small, Stirolab, Sežana, Slovenia) following the DIN EN 12667 standard. The instrument was calibrated using a NIST Standard Reference Material® 1450 d. The typical accuracy of the HFM is ± 1%. The mean temperature of the sample during the measurement was 10 °C, with the upper and lower plate temperatures set at 5 °C and 15 °C, respectively.

The pore size distribution was analyzed based on a magnified cross-section image of the sample. A transparent foil was then placed over the image, and the pores were manually outlined. The resulting image was scanned and processed using ImageJ 1.53t software.

3. Results

Thermal analysis coupled with evolved gas analysis of the sample 12WG-2B-2Al-2K with 12 wt% WG in Ar and air atmosphere revealed important differences (Figure 1). A mass loss (black curves, Figure 1) in both atmospheres occurs practically over the whole temperature range and is related to the release of H_2O and CO_2. A major part of the mass loss is related to a gradual release of water bounded in the water glass [14]. Two peaks in the release of water are located at 120 and 320 °C, while water vapor is present in the evolved gases up to 600 °C.

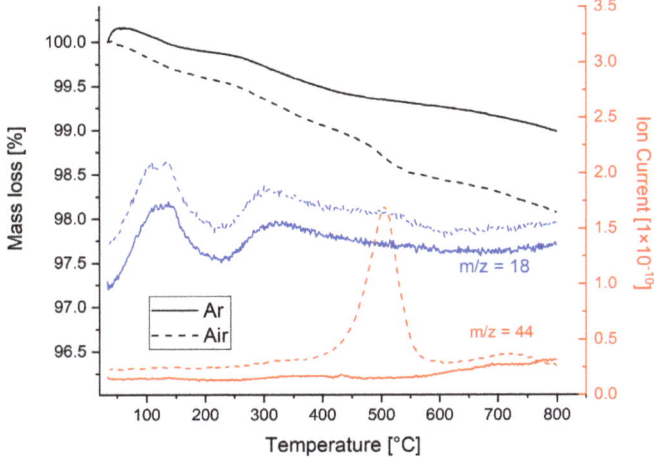

Figure 1. Thermal analysis coupled with evolved gas analysis of the sample.

The main mass loss related to CO_2 release is visible in the sample foamed in an air atmosphere peaking at 500 °C. This is related to the premature oxidation of the added carbon to oxygen present in the atmosphere, which is unwanted [17]. In the sample processed in the Ar atmosphere, a small, i.e., negligible, mass loss is observed at 440 °C. The samples analyzed were in powder form, so sharp peaks of gas release above the sintering temperature are not present. Despite that, in both samples, the signal of CO_2 is observed at temperatures above 600 °C. This is related to (i) the presence of carbon, which is protected from the atmosphere by the added WG and reacts with oxygen from the glass, and (ii) the decomposition of carbonates, which are formed when a wet mixture with WG is exposed to the air atmosphere [15]. Both sources of CO_2 are present in the sample foamed in the Ar atmosphere, while the second one is predominantly present in the sample processed in the air atmosphere. From the mass loss occurring at 500 °C in the air atmosphere, accompanied by the large CO_2 peak, we calculate that around 80% of carbon is burned out. For the sample tested in the Ar atmosphere, the CO_2 signal increases from 600 to 800 °C, while in the sample tested in the air atmosphere, the CO_2 signal peaks at 720 °C and then decreases gradually, indicating that the source of CO_2 is diminishing.

The XRD patterns of the samples foamed in Ar and air atmospheres are shown in Figure 2. The samples processed in the Ar atmosphere contain a higher amount of crystalline phase than the samples foamed in the air atmosphere. The exact mechanism behind this is not known; however, it is most likely related to the presence of carbon, which is higher in the sample foamed in the Ar atmosphere. The carbon binds the oxygen from the glass, thereby changing the oxidation state of the glass, which influences the crystallization processes. In comparison to the samples without added WG [11], the crystalline phase content is lower. In general, the samples with 24 wt% of WG have a significantly lower crystalline content and are very similar in both atmospheres (Figure 2; note: only the XRD pattern of the sample with all additives and 24 wt% WG is shown). For the samples with 24 wt% of WG, the samples processed in both atmospheres exhibit practically identical XRD patterns, and there are negligible differences, e.g., a peak at 36° 2θ foamed in Ar.

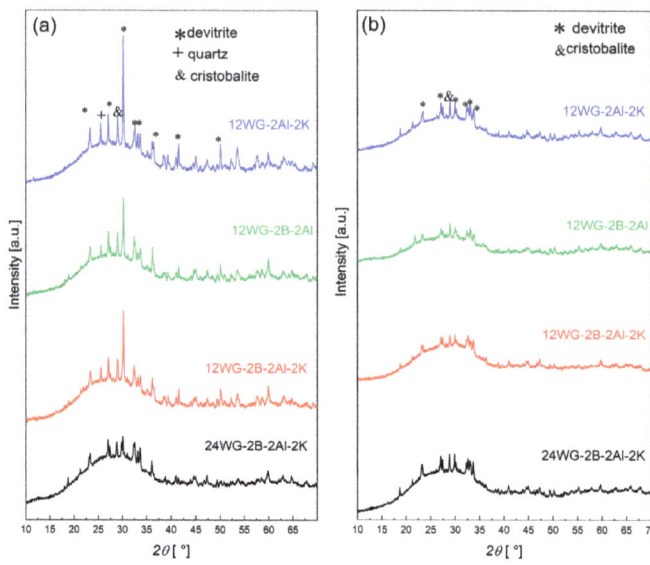

Figure 2. XRD patterns of the samples foamed in (**a**) Ar and (**b**) air atmosphere at 865 °C.

The densities of the samples with 12 wt% of WG foamed in Ar atmosphere are in the range of 143–157 kg m^{-3}, while the samples foamed in air exhibit higher densities in the range of 224–290 kg m^{-3}. In general, the samples with both phosphate additions have a lower density. Closed porosity is the highest in the samples containing all additives. When the amount of added WG is increased to 24 wt%, the densities of the samples foamed in Ar remain at the same level, while for the samples foamed in air atmosphere, the densities decrease significantly. This decrease can be related to a higher amount of carbon being protected by the added WG. The open porosity, however, increases in almost all samples with a higher addition of WG. In Ar, the lowest density obtained is 130 kg m^{-3}, while in the air atmosphere it is 156 kg m^{-3}, in both cases with 24 wt% WG (Table 2).

Table 2. The density, closed porosity, and total porosity of the samples foamed for 10 min at 865 °C in Ar and air atmosphere.

	Ar			Air		
Sample	ρ_{app} (kg m^{-3})	Closed Porosity (%)	Total Porosity (%)	ρ_{app} (kg m^{-3})	Closed Porosity (%)	Total Porosity (%)
12WG-2Al-2K	143	90	94	224	83	91
12WG-2B-2Al	157	90	94	290	85	88
12WG-2B-2Al-2K	143	90	94	211	91	93
24WG-2Al-2K	130	50	95	195	73	92
24WG-2B-2Al	160	88	93	161	58	93
24WG-2B-2Al-2K	142	79	95	156	87	94

4. Large Samples

Large samples from the compositions containing all additives and 12 or 24 wt% WG were prepared to properly evaluate the achievable densities and thermal conductivities. The pore structure of the large samples is shown in Figure 3. The samples foamed in the Ar atmosphere exhibit a more homogeneous pore structure than the samples foamed in the air atmosphere, which also contain larger pores. These differences indicate that the foaming mechanism changes when the atmosphere changes. In the Ar atmosphere, carbon is fully protected and remains in the sintered sample. Oxidation of carbon, with the chemically bounded oxygen in the glass and manganese oxide [17], as well as the decomposition of carbonates formed on addition of WG [15], contribute to the foaming process. The color of the samples is gray (to better understand the color references mentioned, readers are directed to consult the online version of the article). In the air atmosphere, only a small part of carbon remains in the sample, although it is expected for the amount to increase with an increasing sample size and WG content. Thus, the foaming is in major part related to the release of oxygen from the manganese oxide and the decomposition of carbonates formed in the wet foaming mixture with WG. The color of the samples is more purple than gray, indicating that the majority of manganese in the foamed sample is present as Mn^{3+} [18]. The larger pore sizes of these samples foamed in the air indicate that the foaming time could be shortened in order to obtain a sample with smaller pores.

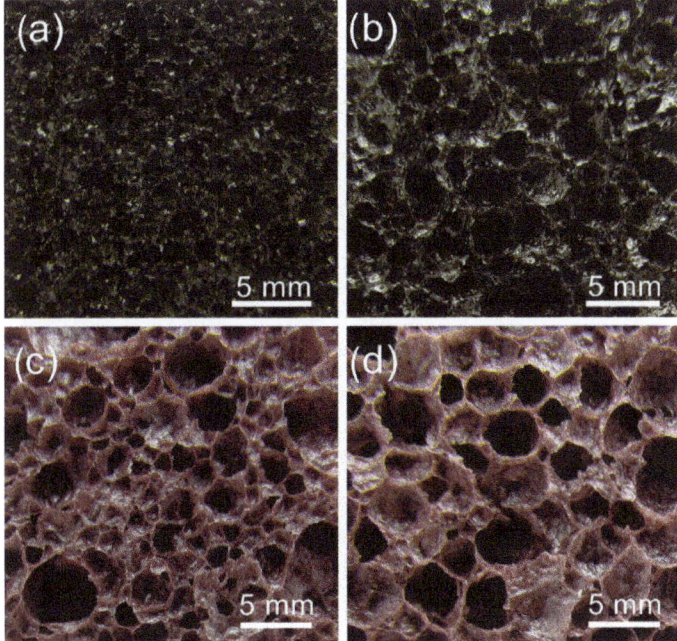

Figure 3. Microstructure of the foamed large samples: (**a**) 12WG-2B-2Al-2K at 865 °C in Ar, (**b**) 24WG-2B-2Al-2K at 865 °C in Ar, (**c**) 12WG-2B-2Al-2K at 880 °C in Air and (**d**) 24WG-2B-2Al-2K at 855 °C in air.

The densities of the large samples with 12 wt% of WG are similar to those of the small samples; however, the densities of the large samples foamed with 24 wt% of WG are significantly lower, Table 3. This is related to the larger size of the sample, where more released water and carbon can stay in the sample, since the diffusion path of the gases in the compacted powder becomes longer. Similarly, the higher content of remaining hydroxyl groups and water in the large samples can contribute to pore coalescence through the decrease of viscosity. The 12 wt% addition is not adequate for triggering such differences. Additionally, the viscosity decreases with an increasing WG content due to the increase of sodium content in the glass [14].

Table 3. Density, porosity and thermal conductivity of the samples foamed for 20 min in Ar and air atmosphere.

Sample	Ar			Air	
	12WG-2B-2Al-2K		24WG-2B-2Al-2K	12WG-2B-2Al-2K	24WG-2B-2Al-2K
Temperature (°C)	865	880	865	880	855
ρ_{app} (kg m^{-3})	147	138	118	147	123
Closed Porosity (%)	80	38	20	52	36
Thermal conductivity (mW m^{-1} K^{-1})	59	57	53	53	54

The thermal conductivities are in the range of 53–66 mW m^{-1} K^{-1}, which is slightly lower than in the samples without WG addition [11] and in the range of commercial foamed glass produced in air atmosphere [19], Table 3. The samples prepared from the composition with 12 wt% WG foamed in Ar and air atmospheres have the same density, but the thermal conductivity of the sample foamed in air is much lower, despite the higher open porosity. Although not measured on this set of samples, based on our previous reports [19,20], we presume that CO_2 is present in the closed pores (thermal conductivity 16 mW m^{-1} K^{-1}), while air with a higher thermal conductivity (25 mW m^{-1} K^{-1}) fills the open pores. From these results, we can conclude that the difference in the thermal conductivity is in major part related to the contribution of the solid (glassy) phase. The crystallinity of the samples with 12 wt% WG foamed in Ar is higher than of those foamed in air, which means that the contribution of the solid conductivity is higher. The sample with 24 wt% WG foamed in Ar has a similar crystalline content to the sample with 12 wt% WG foamed in air, and its thermal conductivity is lower, despite the higher open-porous content. This result again shows that the prevention of crystallization is an important parameter in the production of foamed glass with improved insulation properties [19,21].

Open porosity is another critical parameter. In this set of samples, only the sample with 12 wt% of WG foamed in Ar at 865 °C has a closed porosity of 80%, while the other samples have much lower values of closed porosity. Open porosity also contributes to a higher thermal conductivity [19]. If the samples foamed in the air atmosphere would be fully closed-porous and the pores were filled with CO_2, the thermal conductivity would decrease by 4–6 mW m^{-1} K^{-1}. The achieved values would then be below the commercial reference of 52 mW m^{-1} K^{-1} [22]. In relation to open porosity, one could also expect that the thermal conduction through the solid phase would decrease, since all the material would be placed in the struts (no walls), which decreases the solid conduction [17]. However, the distribution of the solid mass between struts and walls in foamed glass is not so extreme as to trigger such an effect. Moreover, foamed glass with open porosity is not appropriate for thermal insulation in conventional applications on the outer surface of a wall [23] due to the danger of water penetration. Such foamed glass can only be used as acoustic or thermal insulation in the interior of buildings.

The presented results show that bottle glass composition has a lower potential for use in foamed glass boards production. The main issue is related to crystallization, which occurs during foaming and triggers an increase of open porosity and thermal conductivity of solid phase. A new approach is needed in order to be able to use waste bottle glass in the production of foamed glass boards, preferably in the air atmosphere, for thermal insulation applications. One possibility is to change the foaming agent(s), but also to prepare a mixture with flat glass [9]. Until then, this source of waste glass can effectively be used in the production of foamed glass gravel, which typically uses SiC as the foaming agent [24].

5. Conclusions

We investigated the use of water glass in the foaming of waste bottle glass with the carbon–manganese-oxide foaming couple in Ar and air atmosphere. The results show that with an increased addition of WG, the crystallinity and the thermal conductivity decrease in comparison to the samples without WG addition. However, the remaining crystallization greatly influences the properties of the prepared foams, resulting in open porosity and higher thermal conductivities in comparison to amorphous foams in the literature. The DTA analysis revealed that 12 wt% of added WG partly protects the carbon from premature oxidation by the air from the atmosphere at around 500 °C. However, as indicated by the differences in the large samples processed in Ar and air atmospheres, the foaming mechanism differs greatly, and it was not possible to obtain a closed porous sample in air atmosphere. With the WG addition, it was not possible to obtain a foam of proper quality from the waste bottle glass in the air atmosphere. The lowest obtained foam density in the air atmosphere was 123 kg m^{-3}, while the lowest thermal conductivity was 53 mW m^{-1} K^{-1}.

Author Contributions: Conceptualization, J.K. and S.S.; methodology, J.K. and S.S.; formal analysis, U.H.; investigation, S.S.; writing—original draft preparation, S.S. and J.K.; writing—review and editing, S.S., J.K. and U.H.; supervision, J.K.; funding acquisition, M.S. All authors have read and agreed to the published version of the manuscript.

Funding: Slovenian Ministry of Education, Science and Sport, and the European Regional Development Fund (grant number C3330-19-952051) and Slovenian Research Agency (grant number L2-9221).

Institutional Review Board Statement: Not applicable.

Informed Consent Statement: Not applicable.

Data Availability Statement: No further data available.

Acknowledgments: The authors want to acknowledge the support from the Slovenian Ministry of Education, Science and Sport, and the European Regional Development Fund (grant number C3330-19-952051), as well as the Slovenian Research Agency (grant number L2-9221).

Conflicts of Interest: The authors declare no conflict of interest.

References

1. Rodrigues, C.; König, J.; Freire, F. Prospective life cycle assessment of a novel building system with improved foam glass incorporating high recycled content. *Sustain. Prod. Consum.* **2023**, *36*, 161–170. [CrossRef]
2. Hill, C.; Norton, A.; Dibdiakova, J. A comparison of the environmental impacts of different categories of insulation materials. *Energy Build.* **2018**, *162*, 12–20. [CrossRef]
3. Pittsburgh Corning Europe. *Foamglas Industrial Insulation Handbook*; Pittsburgh Corning Europe: Waterlo, Belgium, 1992.
4. Llaudis, A.S.; Tari, M.J.O.; Ten, F.J.G.; Bernardo, E.; Colombo, P. Foaming of flat glass cullet using Si$_3$N$_4$ and MnO$_2$ powders. *Ceram. Int.* **2009**, *35*, 1953–1959. [CrossRef]
5. Steiner, A.C. *Foam Glass Production from Vitrified Municipal Waste Fly Ashes Door*; Eindhoven University Press: Eindhoven, The Netherlands, 2006.
6. García-Ten, J.; Saburit, A.; Orts, M.J.; Bernardo, E.; Colombo, P. Glass foams from oxidation/reduction reactions using SiC, Si$_3$N$_4$ and AlN powders. *Glass Technol. Eur. J. Glass Sci. Technol. Part A* **2011**, *52*, 103–110.
7. Bayer, G. Foaming of borosilicate glasses by chemical reactions in the temperature range 950–1150 °C. *J. Non-Cryst. Solids* **1980**, *38–39*, 855–860. [CrossRef]
8. Scarinci, G.; Brusatin, G.; Bernardo, E. Glass Foams. *Cell. Ceram. Struct. Manuf. Prop. Appl.* **2005**, *2*, 158–176. [CrossRef]
9. König, J.; Petersen, R.R.; Iversen, N.; Yue, Y. Suppressing the effect of cullet composition on the formation and properties of foamed glass. *Ceram. Int.* **2018**, *44*, 11143–11150. [CrossRef]
10. Smiljanić, S.; Hribar, U.; Spreitzer, M.; König, J. Influence of additives on the crystallization and thermal conductivity of container glass cullet for foamed glass preparation. *Ceram. Int.* **2021**, *47*, 32867–32873. [CrossRef]
11. Smiljanić, S.; Spreitzer, M.; König, J. Application of the container waste glass in foamed glass production. *Open Ceram.* **2023**, *14*, 100339. [CrossRef]
12. Hesky, D.; Aneziris, C.G.; Groß, U.; Horn, A. Water and waterglass mixtures for foam glass production. *Ceram. Int.* **2015**, *41*, 12604–12613. [CrossRef]
13. Méar, F.O.; Podor, R.; Lautru, J.; Genty, S.; Lebullenger, R. Effect of the process atmosphere on glass foam synthesis: A high-temperature environmental scanning electron microscopy (HT-ESEM) study. *Ceram. Int.* **2021**, *47*, 26042–26049. [CrossRef]

14. Hribar, U.; Østergaard, M.B.; Iversen, N.; Spreitzer, M.; König, J. The mechanism of glass foaming with water glass. *J. Non-Cryst. Solids* **2023**, *600*, 122025. [CrossRef]
15. Hribar, U.; Spreitzer, M.; König, J. Applicability of water glass for the transfer of the glass-foaming process from controlled to air atmosphere. *J. Clean. Prod.* **2020**, *282*, 125428. [CrossRef]
16. Østergaard, M.B.; Petersen, R.R.; König, J.; Yue, Y. Effect of alkali phosphate content on foaming of CRT panel glass using Mn3O4 and carbon as foaming agents. *J. Non-Cryst. Solids* **2018**, *482*, 217–222. [CrossRef]
17. König, J.; Petersen, R.R.; Yue, Y.; Suvorov, D. Gas-releasing reactions in foam-glass formation using carbon and $MnxOy$ as the foaming agents. *Ceram. Int.* **2017**, *43*, 4638–4646. [CrossRef]
18. Iyel, A.; Oktem, D.; Akmaz, F. Parameters Affecting the Color Mechanism of Manganese Containing Colored Glasses. *J. Chem. Chem. Eng.* **2014**, *9*, 849–858.
19. König, J.; Lopez-Gil, A.; Cimavilla-Roman, P.; Rodriguez-Perez, M.A.; Petersen, R.R.; Østergaard, M.B.; Iversen, N.; Yue, Y.; Spreitzer, M. Synthesis and properties of open- and closed-porous foamed glass with a low density. *Constr. Build. Mater.* **2020**, *247*, 118574. [CrossRef]
20. PCimavilla-Román, P.; Villafañe-Calvo, J.; López-Gil, A.; König, J.; Rodríguez-Perez, M. Modelling of the mechanisms of heat transfer in recycled glass foams. *Constr. Build. Mater.* **2021**, *274*, 122000. [CrossRef]
21. Østergaard, M.B.; Petersen, R.R.; König, J.; Johra, H.; Yue, Y. Influence of foaming agents on solid thermal conductivity of foam glasses prepared from CRT panel glass. *J. Non-Cryst. Solids* **2017**, *465*, 59–64. [CrossRef]
22. Programme, D.; Declaration, P. GLAPOR cellular glass GLAPOR Werk Mitterteich GmbH. *Mitterteich Ger.* **2017**, 1–8. Available online: https://www.glapor.de/ (accessed on 30 May 2023).
23. *EN 13167:2012+A1:2015*; Thermal Insulation Products for Buildings—Factory Made Cellular Glass (CG). European Committee for Standardization: Brussels, Belgium, 2015. Available online: https://standards.iteh.ai/catalog/standards/cen/e2f42873-03de-45 7f-9fc8-6541f8541f64/en-13167-2012a1-2015 (accessed on 28 July 2022).
24. Hibbert, M. *Understanding the Production and Use of Foam Glass Gravel across Europe and Opportunities in the UK*; Chartered Institution of Wastes Management: Northampton, UK, 2016.

Disclaimer/Publisher's Note: The statements, opinions and data contained in all publications are solely those of the individual author(s) and contributor(s) and not of MDPI and/or the editor(s). MDPI and/or the editor(s) disclaim responsibility for any injury to people or property resulting from any ideas, methods, instructions or products referred to in the content.

Article

Effect of a Phosphorus Additive on Luminescent and Scintillation Properties of Ceramics GYAGG:Ce

Lydia V. Ermakova [1], Valentina G. Smyslova [1], Valery V. Dubov [1], Daria E. Kuznetsova [1], Maria S. Malozovskaya [1], Rasim R. Saifutyarov [1], Petr V. Karpyuk [1], Petr S. Sokolov [1,*], Ilia Yu. Komendo [1], Aliaksei G. Bondarau [2], Vitaly A. Mechinsky [2] and Mikhail V. Korzhik [2]

[1] National Research Center "Kurchatov Institute", 123098 Moscow, Russia; ermakova.lydia@gmail.com (L.V.E.); smyslovavg@gmail.com (V.G.S.); valery_dubov@mail.ru (V.V.D.); daria_kyznecova@inbox.ru (D.E.K.); puellamay@mail.ru (M.S.M.); rosya01@gmail.com (R.R.S.); silancedie@mail.ru (P.V.K.); i.comendo@gmail.com (I.Y.K.)

[2] Institute for Nuclear Problems, Belarus State University, 220030 Minsk, Belarus; alesonep@gmail.com (A.G.B.); vitaly.mechinsky@cern.ch (V.A.M.); mikhail.korjik@cern.ch (M.V.K.)

* Correspondence: sokolov-petr@yandex.ru

Abstract: The production of scintillating ceramics can require the utilization of the phosphorus compounds at certain stages of 3D-printing, such as vat polymerization, applied for the formation of green bodies before sintering. The effect of phosphorus additive on the microstructure, optical, and scintillation parameters of $Gd_{1.494}Y_{1.494}Ce_{0.012}Al_2Ga_3O_{12}$ (GYAGG:Ce) ceramics obtained by pressureless sintering at 1650 °C in an oxygen atmosphere was investigated for the first time. Phosphorus was introduced in the form of $NH_4H_2PO_4$ into the initial hydroxycarbonate precipitate in a wide concentration range (from 0 to 0.6 wt.%). With increasing of phosphorus concentration, the density and the optical transmittance of garnet ceramics show a decrease, which is caused by an increase in the number of pores and inclusions. The light yield of fast scintillation, which is caused by Ce^{3+} ions, was found to be affected by the phosphorus additive as well. Moreover, an increase in phosphorescence intensity was recognized.

Keywords: ceramics; garnet; phosphorus; luminescence; phosphorescence; stereolithography

1. Introduction

Garnet structure oxides doped with lanthanides are a group of widely used luminescence, laser, and scintillation materials [1–15]. $Y_3Al_5O_{12}$:Ce (YAG:Ce) and $Lu_3Al_5O_{12}$:Ce (LuAG:Ce) are well-established scintillators, which are widely used in radiation detection applications in science and industry. Nevertheless, recently, along with binary compositions, such as YAG:Ce and LuAG:Ce, multicationic garnets have been actively studied [1–3,5–9]. The garnet matrix has been developed to become more complex; yttrium is partially or completely replaced by gadolinium or a Gd/Lu mixture in different ratios; and aluminum is partially replaced by gallium [8–11]. In addition, garnets doped with other lanthanides, or their combination are being actively studied [1,5–7,12]. Compositionally disordered garnet structure compounds with a general formula $(Gd,Y,Lu)_3(Al,Ga)_5O_{12}$, doped with rare earth activator(s) and, facultatively co-doped with other element(s) became in the focus of the research due to a unique set of features: high chemical stability, high density, high effective atomic number, high light yield, fast scintillation kinetics, etc. [6,8–10,15–17]. It can be produced in the form of both single crystals [9,10,16–18] and translucent [15] or transparent ceramics [8,10,13,19].

Luminescent ceramics have some advantages in comparison to single crystals of the same structure and composition. Ceramics can be produced more cheaply, potentially any shape, almost any size and/or composition. Also, new structures are accessible due to the versatility of ceramics, e.g., composites [20–23]. At the same time, the major functional properties of highly transparent ceramics could be due to single crystals.

Various additive manufacturing techniques were applied to produce transparent [22,23] or translucent [24,25] garnet ceramics. Material jetting [22] and direct ink write [23] methods were used for the formation of green body further used for the sintering. YAG:Yb/YAG:Nd and YAG:Er/YAG:Lu transparent all-ceramic composites of disk-like [22] or rod-like [23] shapes were obtained by vacuum/air sintering with subsequent hot isostatic pressing. Also new methods for sintering garnet ceramics under electron-beam [26] and laser [27] irradiation seem promising. Nevertheless, one of the most frequently developed methods of 3D-printing suitable for mass applications, is stereolithography. It provides one of the best spatial resolutions, a smooth surface of printed objects with an acceptable building speed and a possibility to use the pressureless sintering [24,25].

Obviously, the key properties of garnet ceramics depend on the chemical composition and the perfection of the oxide matrix [12,14,15], the chosen activators [28–31], technological factors [19,30–33], synthesis conditions [8] and post-processing treatment [31]. Another important factor is the nature and a concentration of impurities which come in the ceramics at different stages of the technology. There are a plenty of publications describing the influence of cations of various metals, such as alkali [32] and alkaline earth elements [9,16,17,31,33,34], elements of the third [18,20] and fourth [34] groups. The effect of silicon [22,23,33–35] and boron [33,36] additives has been also well clarified. Silicon is a widely used sintering additive [[33] and refs therein], which is utilized to prepare transparent or translucent garnet ceramics. The disadvantage of using such sintering additives is their potential negative effects on the luminescent and scintillation properties of the resulting ceramics [32,33,36].

Research on the influence of other non-metals, such as nitrogen, is much less described [35]. Phosphorus is a typical non-metal element, a neighbor of nitrogen and silicon in the periodic table. However, to the best of our knowledge, the effect of phosphorus additives on garnet ceramics has not been practically studied before. Only single article has been recently published, where YPO_4/YAG:Ce nanocomposites were purposefully synthesized and studied in details [37]. At the same time, it is well known that phosphorus-containing dispersants can be used in the preparation of slurries in ceramic technology [38,39], including slurries for 3D-printing. Phosphoric acid ester derivatives have high wetting characteristics for surface oxide powders. It allows to reach a high loading of slurries with acceptable rheological properties [38,39]. The typical value of specific surface area (SSA) is from 45 to 60 m^2/g and from 3 to 12 m^2/g, for garnet oxide powders annealed at 850 and 1300 °C [24,25], respectively. The content of the dispersant in slurry is usually proportional to the SSA of the ceramic powder and can be reached up to 3 mg for each m^2 [24,25]. According to our preliminary study, the phosphorus content in commercially available dispersants is about 4 wt.%.

Moreover, UV photocurable slurries with ceramic particles for stereolithography 3D printing may contain other phosphorus compounds, like UV photoinitiators of radical polymerization of the class of phosphine oxides (BAPO, TPO, TPO-L, etc.). The typical content of such photoinitiators is ~1.0 wt.% based on the weight of acrylate monomers [24,25]. Thus, the potential content of phosphorus in the slurry can be quite high value (1–7 mg for each g of powder or 1000–7000 ppm). The high sintering temperature could induce the partial volatilization of phosphorus, which may result in loose microstructure of garnet ceramics. The formation of impurity phases is also very possible.

Here, we report for the first time an effect of phosphorus impurity on the major functional properties of doped GYAGG:Ce garnet scintillation ceramics. The key properties of the sintered ceramics were correlated with the amount of phosphorus.

2. Materials and Methods

2.1. Synthesis of Initial Powders

Starting powder was fabricated by co-precipitation method [5–7,11–13,33]. High purity commercially available chemical reagents such as Gd_2O_3 (99.995%), Y_2O_3 (99.995%),

AlOOH (99.998%), Ga (99.999%) and Ce(NO$_3$)$_3$ (99.95%) were used as raw materials to prepare nitrate solutions. The solutions were mixed in the required proportions to obtain composition Gd$_{1.494}$Y$_{1.494}$Ce$_{0.012}$Al$_2$Ga$_3$O$_{12}$ and diluted to obtain the total Me^{3+} ion concentration of 0.5 mol/L. Next, the mixed solution was slowly added to the precipitant—a solution of ammonium bicarbonate NH$_4$HCO$_3$ (99.95%) with a concentration of 1.5 mol/L—under constant stirring with an overhead stirrer. The hydroxocarbonate precipitate was filtered, washed with high-purity isopropyl alcohol (IPA)–distilled water mixture a few times, and dried at 80 °C in an air-ventilated oven for 8 h. Further, the precipitate was divided into four equal parts. One part was used as a reference (untreated) sample, the other three parts were utilized to enhance the phosphorus content.

The NH$_4$H$_2$PO$_4$ (99.5%) was used as a source of phosphorus, the details of introducing are described elsewhere [33]. Three IPA-water solutions with different concentration of phosphorus were prepared. Weighed portion of the precipitate was added in each solution. These suspensions were stirred for a day, then dried at 80 °C, and samples were taken for elemental analysis. The motivation of choosing this substance as a source of phosphorus is presented in the Supplementary.

Finally, all four precursors with different phosphorus content were placed in corundum crucibles with caps and calcined together in a muffle furnace at 850 °C for 2 h to form the garnet phase. During annealing, the precipitate showed a weight loss of about 29%. Afterward, the oxide powders were milled in a planetary ball mill with alumina jars and beads to get a median particle size (d_{50}) of 1.5–1.8 µm according to laser diffraction measurements. The grinding conditions were identical for all compositions. Milling media was IPA, rotation speed was 300 rpm; grinding time was 30 min, weight ratio of IPA:powder:beads was 2:1:2. After milling the slurries were dried at 80 °C and sieved through a 100-µm meshes. Samples were taken for elemental analysis. As the result, four samples: nominally pure (#0) and, loaded with phosphorus (#1–3) were produced. The stages of their production and characterization methods are described below.

2.2. Characterization of Initial Powders

Particle size distributions were measured using laser diffraction on a MasterSizer 2000 (Malvern, PA, USA) with a water-filled dispersing unit Hydro G. The specific surface area (SSA) and pore volume of the powders were determined according to the capillary nitrogen condensation method using BET and BJH models on NOVAtouch NT LX (Quantachrome Instruments, New York, NY, USA). The phase compositions of the oxide powders were examined using X-ray powder diffraction on a D2 Phaser (Bruker, Billerica, MA, USA) with CuK$\alpha_{1,2}$ radiation.

Elemental analysis of precipitates and calcined powders was carried out via iCAP 6300 duo (Thermo Scientific, Waltham, MA, USA) spectrometer by the ICP AES method. Before the measurement, the powders are dissolved in a mixture of ultra-pure nitric and hydrochloric acids at temperature of 100 °C using a HotBlock (Environmental express, Ocala, FL, USA) equipment.

2.3. Ceramics Fabrication

Green bodies were prepared by uniaxial pressing at 64 MPa into 1.5 mm-thick pellets of 20 mm in diameter. The typical green density was about 35% of the theoretical density of a single-crystal GYAGG:Ce (6.0 g/cm^3). Then pellets were sintered at 1650 °C 2 h in an oxygen atmosphere by using tube furnace.

Finally, surface of the ceramic samples was grinded with silicon carbide abrasive papers and then polished with 0.5 µm and 0.1 µm diamond polishing pastes. The thickness of the ceramic samples was ~1 mm. The polished samples intended for scanning electron microscopy were additionally thermally etched for 10 min at 1200 °C to reveal grain boundaries.

2.4. Characterization of Ceramic Samples

The apparent density of the ceramic samples was measured using Archimedes' method in Lotoxane at room temperature. The uncertainty in this measurement was about 0.5%.

Ceramic microstructure was studied using a Jeol JSM-7100F (JEOL, Tokyo, Japan) scanning electron microscope (SEM). SEM images were obtained in secondary electrons and backscattered electrons modes. Platinum sputter-coating was used to ensure electrical conductivity of surface ceramic sample. Local chemical compositions were estimated using energy-dispersive X-ray spectroscopy (EDX) via X-Max 50 (Oxford Instruments, London, UK) attachment. Processing of the SEM images to determine the average grain sizes and estimate of the number of inclusions of ceramics was carried out using ImageJ software.

The full transmittance of the ceramic samples in the visible region of the spectrum (400–700 nm) was determined on an Specord Plus spectrophotometer (Analytik Jena, Jena, Germany) equipped with an integrating sphere. The photoluminescence (PL) spectra of the ceramic samples were measured on a Fluorat-02 Panorama spectrofluorimeter (Lumex, Moscow, Russia) with a xenon lamp excitation at room temperature.

The photoluminescence kinetics were studied on a FluoTime 250 luminescence spectrometer (PicoQuant, Berlin, Germany) using a pulsed LED excitation source with a wavelength of 340 nm and a pulse width of 200 ps, corresponding to excitation of the $4f \rightarrow 5d_1$ interconfigurational transition of Ce^{3+} ions.

The light output (LO) of the samples was measured with a ^{137}Cs (662 keV) source by collecting the pulse height spectra with the XP2020 photomultiplier readout. Incident γ-quanta interact with a whole volume of the sample, so the position of the γ-quanta photo-absorption peak in the spectra is affected by the scattering of the scintillation light. Therefore, the light output is smaller than the light yield (LY) due to the reduced light collection factor in the translucent sample.

A thin layer of the sample, not more than 10 μm, absorbs α-particles in the material. Due to this reason, measurement with α-particles in a 45° geometry [40] provides a light yield of scintillation practically from the surface of the sample, which is not affected by scattering. An α-particle source (~5.5 MeV, ^{241}Am) was used to collect the pulse height spectra with the XP2020 photomultiplier readout. A YAG:Ce single-crystal with ground surfaces to mimic ceramics with a light yield of 4100 ph/MeV under α-particles excitation and 25,000 ph/MeV under γ-quanta excitation was used in these measurements as a reference.

3. Results

The results of quantitative elemental analysis of the phosphorus content in the initial powders are presented in Table 1. Sample # 1 contains approximately Ce 1:1 P (mole ratio).

Table 1. The measurement contents of phosphorus in initial powders (wt.%) [1].

Sample #	Hydroxocarbonate Precipitates	Oxide Powders Calcined at 850 °C
0	-	-
1	0.027	0.040
2	0.114	0.156
3	0.456	0.623

[1] According to elemental analysis, the content of cerium in the hydroxocarbonate precipitates and powders calcined at 850 °C is 0.139(1) and 0.191(1) wt.%, respectively, in good agreement with to the expected chemical composition.

It is known [41,42] that ammonium dihydrogen phosphate completely decomposes into gaseous products already at temperatures of about 550 °C. In the same time, based on the results of elemental analysis, we do not observe significant loss (volatilization) of phosphorus. One can assume that as-synthesized ReE (Y, Gd, Ce) oxides may easy react with $NH_4H_2PO_4$ [37,43,44] at relatively low temperature and form the refractory and extremely stable phosphates [37,45,46].

The BET specific surface area and the BJH porosity for garnet powder calcined at 850 °C were 55 m^2/g and 0.3 cm^3/g, respectively. According to X-ray diffraction analysis all initial powders had a garnet crystal structure of Ia-$3d$ with lattice parameter a = 12.232(5) Å, in good agreement with the literature [7].

Diffractograms of ceramics obtained from sample # 2 and # 3 contain a few additional weak lines, which can be attributed to $(Y,Gd)PO_4$ with a tetragonal ($I4_1/amd$) xenotime structure (Figure S1). Lines of monazite-type phosphates (typical for pure $GdPO_4$) are not observed. The resulting ceramic samples had a high density from 100 to 97.5% (Table 2).

Table 2. The average relative density of GYAGG:Ce ceramic samples (%).

Sample #	0	1	2	3
Relative density	100	99.7	98.8	97.5

Thus, the sinterability and densification of garnet powders with phosphorus are apparently decreased. This behavior can be explained by two factors. Firstly, the presence of an impurity of refractory orthophosphate possibly reduces the sinterability due to high melting point of YPO_4~2150 °C [46], which is higher than the melting point of YAG or GAGG compounds [24]. Secondly, the density of yttrium orthophosphate is significantly lower than the density of GYAGG:Ce ceramics (4.27 vs. 6.0 g/cm^3).

The assumptions above are supported by the electron microscopy data (Figure 1). According to SEM analysis, the grains sizes for GYAGG:Ce ceramics without phosphorus (sample # 0) are up to 7 μm; generally the sample # 0 has a homogeneous microstructure, which is typical for dense garnet ceramics [7,8,13,14,33]. According to the EDX analysis, the element content (Gd 28 wt.%; Y 16 wt.%; Al 6 wt.%; Ga 25 wt.%; O~24 wt.%; Ce 0.2 wt.%) was in good agreement with the synthesized composition. Just few pores and inclusions have been observed. Total amount of inclusions and pores is 0.1(1)% (Table 3).

Table 3. The average grain size of garnet phase, total fraction of inclusions and pores in the GYAGG:Ce ceramic samples.

Sample #	0	1	2	3
Average grain size (μm)	1.90 (1)	3.2 (2)	6.9 (2)	2.8 (2)
Fraction of inclusions + pores (%)	0.1 (1)	0.9 (5)	3.5 (5)	16.3 (9)

All the samples contain a number of pores with different shapes and sizes up to 3–5 μm, which progressively increases in number from 1 to 3 series. The number of impure grains increases with the increase of phosphorus in the samples (Figure 1, Table 3) as well. The chemical composition of these grains is slightly variable, in any case they are enriched in phosphorus (up to 12 wt%) and gadolinium (up to 43 wt.%), and depleted in aluminum (down to 1 wt.%) and gallium (down to 3 wt.%), and the same time the yttrium content reaches the 12 wt.%. So, from the comparison of Figure 1d–f, we can conclude that the the main inclusions in sample #3 are phosphates, in good agreement with our X-ray diffraction data and results from [37].

In samples # 2 and # 3 there are also grains of aluminum-gallium oxide. Earlier, appearance of the $(Al,Ga)_2O_3$ oxide phase was observed when Gd content was below the stoichiometric garnet composition [14]. But in this work, their appearance is explained by the fact that some of the yttrium and gadolinium atoms bind to inert phosphates.

The average grain size of GYAGG:Ce ceramics without phosphorus (sample # 0) is 1.9 μm. An increase in the average grain size with an increase in the phosphorus concentration in the garnet ceramics was found for sample 0, 1 and 2 (Table 3). The larger grains were observed in sample # 2 (Figure 1c). The average grain size for sample # 3 is

2.8 µm. It is possible that relatively small amounts of phosphorus lead to more active grain growth due to increase in the defectiveness of the garnet structure. And in the case of an excess of phosphorus (sample # 3), impurity phases come out in the form of individual crystallites and have less effect on grain growth.

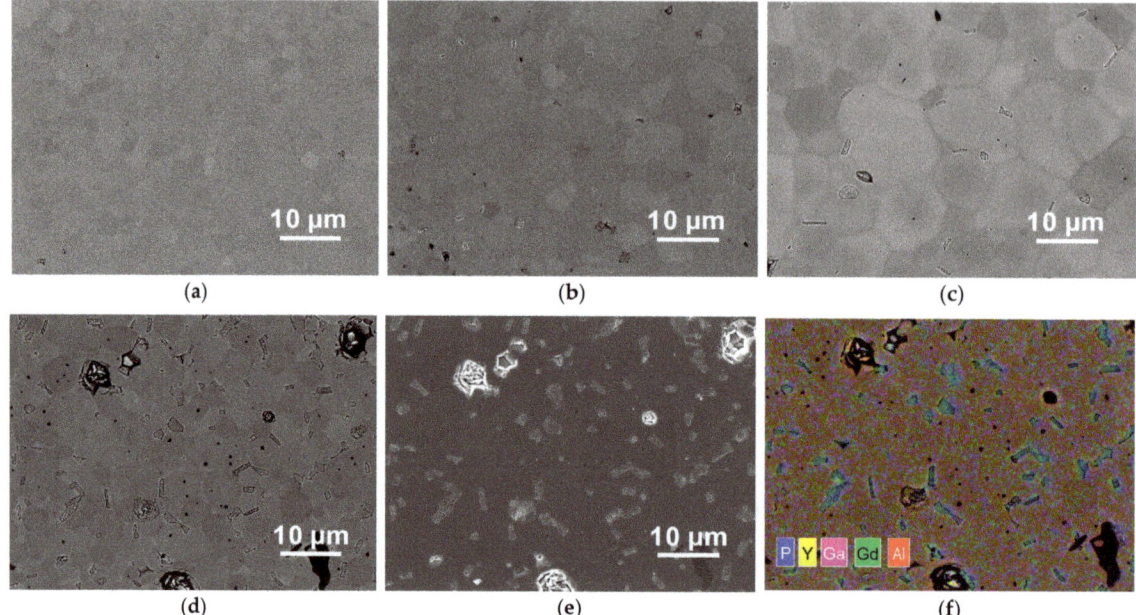

Figure 1. Representative SEM images of mirror polished and thermally etched surfaces of GYAGG:Ce ceramics. (**a**) sample # 0; (**b**) sample # 1; (**c**) sample # 2; (**d**–**f**) sample # 3 (maximum phosphorus content). Sample # 0 is a reference ceramic without phosphorus additives; in other samples, the phosphorus content increases with increasing number. SEM images (**a**–**d**) recorded in backscattered electrons mode and (**e**) recorded in secondary electrons mode; (**f**) element mapping for sample # 3. Scale bar 10 µm.

Optical photographs and additional SEM images (secondary electrons mode) of the ceramic samples are presented in Supplementary as Figure S2 and Figure S3, respectively.

The photoluminescent properties of ceramics are shown in Figure 2. Photoluminescence (PL) and photoexcitation spectra (PLE) of GYAGG:Ce ceramics with phosphorus additives have the characteristic luminescence bands of Ce^{3+} in garnet matrices.

The peak position of the luminescence spectra does not depend on the concentration of phosphorus, while for the excitation spectra, a shift of the excitation band maximum corresponding to the f^1d^0-f^05d^1 transition is observed: from 430 nm for sample # 0 to 450 nm for sample # 3. Worth noting, the luminescence intensity increases with the increasing of phosphorus concentration, passing through a maximum for sample # 2, after which it decreases when passing to sample # 3.

The LO of the ceramic samples correlates with their translucence. Changes in the LO and optical transmission at 520 nm, which correspond to the maximum of the scintillation spectrum, are shown in Figure 3. It is worth stating that there is a clear deterioration of the LO with the transmittance reducing.

Figure 4 shows the pulse height spectra of samples measured upon excitation by alpha particles. The position of the total absorption peak correlates with the scintillation yield of the sample. The positions of the total absorption peaks are as follows: YAG:Ce reference (208 ch.); #0 (331 ch); #1 (306 ch); #2 (261 ch); and #3—no resolved peak at all. Thus, there is a progressive decrease in the LY caused by the fast scintillation of Ce^{3+} ions as the phosphorus concentration increases in the sample. Moreover, scintillation was practically suppressed at the highest phosphorus concentration.

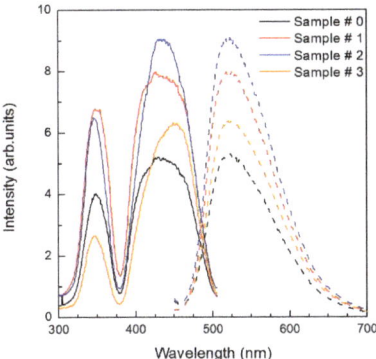

Figure 2. Room temperature measured luminescence (λ_{ex} = 350 nm) and its excitation spectra (λ_{reg} = 520 nm) of GYAGG:Ce ceramics. Number of the series is indicated. Solid lines are PLE spectra; Dashed lines are PL spectra.

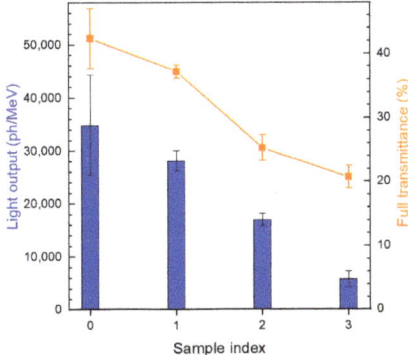

Figure 3. Change of the light output and optical transmittance at 520 nm in the GYAGG:Ce ceramics samples having different concentration of phosphorus. Samples indexes are indicated in Table 1.

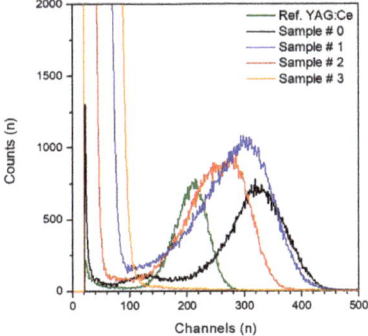

Figure 4. Pulse-height spectra of garnet ceramics and reference YAG:Ce samples measured under α-particles.

The photoluminescence kinetics of the samples from different series are compared in Figure 5. There is a progressive shortening of the initial stage of the kinetics curve with increasing phosphorus concentration, which indicates quenching of the photoluminescence of Ce^{3+} ions. This process contributes to the decrease in scintillation light yield. But this is not the only process of deterioration; most likely, phosphorus creates a deep electron trapping center, which competes with Ce^{3+} ions to catch non-equilibrium carriers and, at its thermal ionization, provides phosphorescence. This suggestion is supported by an increase in the intensity of the plateau in Figure 5b after the fast photoluminescence stage, which indicates a significant increase in the phosphorescence of the emitting light.

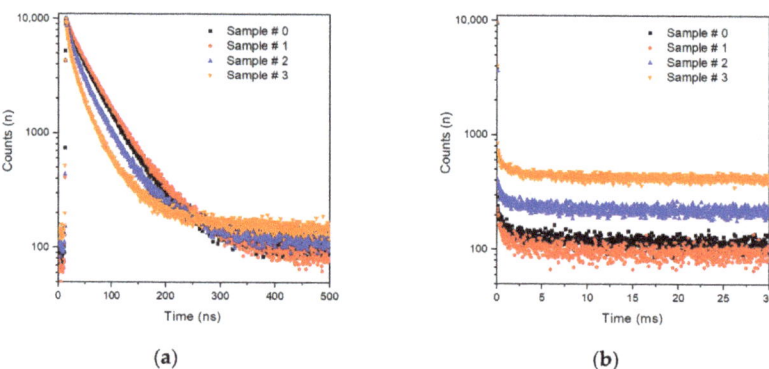

Figure 5. Room temperature PL kinetics of the of GYAGG:Ce ceramic samples at λ_{reg} = 550 nm and excitation λ_{ex} = 340 nm: (**a**)—in a typical time scale applied for Ce^{3+} luminescence kinetics measurements; (**b**)—in a millisecond's scale, which is suitable to observe phosphorescence.

Figure 6 demonstrates the intensity of the persistent luminescence, which intensity is progressively increased in the samples as the phosphorus concentration increases. Persistent luminescence on the second time scale or longer is clearly observed.

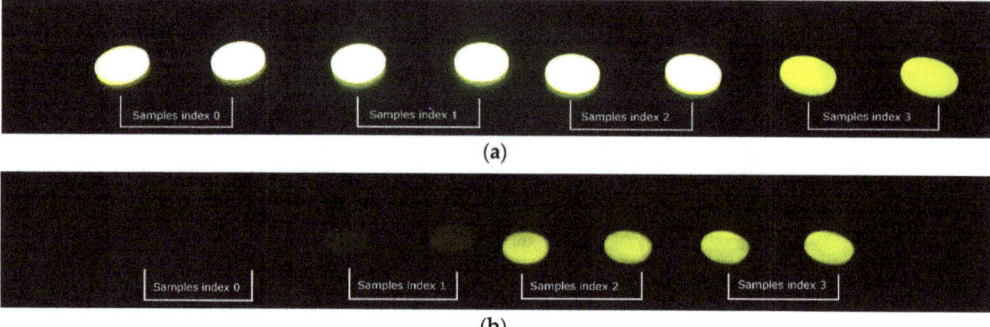

Figure 6. Images of ceramic samples of different series under UV illumination of mercury lamp (λ = 254 nm) (**a**) and 10 s after switch-off the lamp (**b**). Intensity of the persistent luminescence correlates with an increase of the phosphorus in the sample.

4. Discussion

The effect of phosphorus additives on the optical and physical properties of garnet ceramics was found to be quite strong. Apparently, it is due to the relatively high chemical activity of the as-synthesized oxide powders from the hydroxocarbonate precipitate. Phosphorus chemically binds the rare earth elements (Y, Gd, Ce) into inert orthophosphate. As a result, during the process of creating garnet ceramics, a depletion in the concentration of rare earth elements relative to their stoichiometric composition occurs, *viz.*, the phosphates should form second phases (inclusions) in the garnet-type oxides. Recently, the localization of phosphate with respect to cerium luminescence centers in the YAG host has been evaluated by high-resolution scanning transmission electron microscopy and shows distinct YPO_4 and YAG phases in nanocomposite [37]. Based on crystal chemistry, the incorporation of phosphorus into the garnet structure is practically impossible. Indeed, phosphorus contents do not exceed 1 wt% P_2O_5 in garnet minerals [47,48], and are present mostly as inclusions. To the best of our knowledge, in literature there is just one example, when phosphate forms the garnet crystal structure. To achieve this, a very extreme synthesis conditions were required. The $Na_3Al_2(PO_4)_3$ compound with garnet-like crystal structure and its solid solutions were synthesized at high-pressure (>15 GPa) and high-temperature (>1200 °C) conditions [49]. So, under ambient pressure phosphates will form a separate phase(s).

Thus, governing the amount of phosphorus in the initial reagents and throughout the whole process of making ceramics is an important issue. Even a trace concentration of phosphorus in the ceramics results in an increase in phosphorescence. As a result, such parameters of the scintillation material as the afterglow will suffer.

5. Conclusions

For the first time we systematically studied of phase compositions, microstructure, and optical properties of GYAGG:Ce scintillation ceramics with different amounts of phosphorus additives. This is considered to be important, in view of the utilization of the phosphorus chemicals in 3D printing, for precursor densification. Major characteristics of GYAGG:Ce ceramics were found to depend on the amount of phosphorus additives. With increasing phosphorus, the number of defects in the ceramics (pores, secondary phases) increases drastically. Optical transmittance, density, and scintillation yield under alpha- and gamma-excitation are systematically decreased. Phosphorescence intensity shows significant growth as well. All these circumstances require governing the amount of phosphorus in the initial reagents and throughout the whole process of making ceramics.

Supplementary Materials: The following supporting information can be downloaded at: https://www.mdpi.com/article/10.3390/ceramics6030091/s1, Figure S1: X-ray diffraction pattern of GYAGG:Ce ceramics sample # 3; Figure S2: Optical images of GYAGG:Ce ceramics; Figure S3: Representative SEM images (5000×) recorded in secondary electrons mode of mirror polished and thermally etched surfaces of GYAGG:Ce ceramics.

Author Contributions: Conceptualization, P.S.S., I.Y.K. and D.E.K.; methodology, V.G.S., V.V.D. and L.V.E.; software, V.V.D. and P.V.K.; validation, P.S.S. and I.Y.K.; formal analysis, V.V.D.; investigation, L.V.E., V.V.D., P.V.K., R.R.S., M.S.M., A.G.B. and V.A.M.; resources, D.E.K.; data curation, V.V.D.; writing—original draft preparation, P.S.S. and L.V.E.; writing—review and editing, D.E.K., M.V.K. and I.Y.K.; visualization, P.V.K. and P.S.S.; supervision, M.V.K. and D.E.K.; project administration, D.E.K. and P.S.S.; funding acquisition, D.E.K. All authors have read and agreed to the published version of the manuscript.

Funding: This research received no external funding.

Institutional Review Board Statement: No Institutional Review Board Statement is required.

Informed Consent Statement: Not applicable.

Data Availability Statement: No new and additional data are available.

Acknowledgments: Analytical research was conducted using the equipment of the «Research Chemical and Analytical Center NRC, Kurchatov Institute», Shared Research Facilities under project's financial support by the Russian Federation, represented by The Ministry of Science and Higher Education of the Russian Federation, Agreement No. 075-15-2023-370 dd. 22 February 2023. Powder synthesis, optical characteristics' measurements, and ceramics sintering were performed with the support of the grant of the Russian Science Foundation No. 22-13-00172, https://rscf.ru/en/project/22-13-00172/ (accessed on 5 July 2023), in NRC, "Kurchatov institute".

Conflicts of Interest: The authors declare no conflict of interest.

References

1. Ueda, J.; Tanabe, S. Review of luminescent properties of Ce^{3+}-doped garnet phosphors: New insight into the effect of crystal and electronic structure. *Opt. Mater. X* **2019**, *1*, 100018. [CrossRef]
2. Xia, Z.; Meijerink, A. Ce^{3+}-Doped garnet phosphors: Composition modification, luminescence properties and applications. *Chem. Soc. Rev.* **2017**, *46*, 275–299. [CrossRef] [PubMed]
3. Wu, J.L.; Gundiah, G.; Cheetham, A.K. Structure–property correlations in Ce-doped garnet phosphors for use in solid state lighting. *Chem. Phys. Lett.* **2007**, *441*, 250–254. [CrossRef]
4. Palashov, O.V.; Starobor, A.V.; Perevezentsev, E.A.; Snetkov, I.L.; Mironov, E.A.; Yakovlev, A.I.; Balabanov, S.S.; Permin, D.A.; Belyaev, A.V. Thermo-optical studies of laser ceramics. *Materials* **2021**, *14*, 3944. [CrossRef] [PubMed]
5. Korjik, M.; Bondarau, A.; Dosovitskiy, G.; Dubov, V.; Gordienko, K.; Karpuk, P.; Komendo, I.; Kuznetsova, D.; Mechinsky, V.; Pustovarov, V.; et al. Lanthanoid-doped quaternary garnets as phosphors for high brightness cathodoluminescence-based light sources. *Heliyon* **2022**, *8*, E10193. [CrossRef] [PubMed]
6. Korzhik, M.; Abashev, R.; Fedorov, A.; Dosovitskiy, G.; Gordienko, E.; Kamenskikh, I.; Kazlou, D.; Kuznecova, D.; Mechinsky, V.; Pustovarov, V.; et al. Towards effective indirect radioisotope energy converters with bright and radiation hard scintillators of (Gd,Y)$_3$Al$_2$Ga$_3$O$_{12}$ family. *Nucl. Eng. Technol.* **2022**, *54*, 2579–2585. [CrossRef]
7. Korzhik, M.; Borisevich, A.; Fedorov, A.; Gordienko, E.; Karpyuk, P.; Dubov, V.; Sokolov, P.; Mikhlin, A.; Dosovitskiy, G.; Mechinsky, V.; et al. The scintillation mechanisms in Ce and Tb doped (Gd$_x$Y$_{1-x}$)Al$_2$Ga$_3$O$_{12}$ quaternary garnet structure crystalline ceramics. *J. Lumin.* **2021**, *234*, 117933. [CrossRef]
8. Retivov, V.; Dubov, V.; Komendo, I.; Karpyuk, P.; Kuznetsova, D.; Sokolov, P.; Talochka, Y.; Korzhik, M. Compositionally disordered crystalline compounds for next generation of radiation detectors. *Nanomaterials* **2022**, *12*, 4295. [CrossRef]
9. Martinazzoli, L.; Nargelas, S.; Bohacek, P.; Cala, R.; Dusek, M.; Rohlicek, J.; Tamulaitis, G.; Auffray, E.; Nikl, M. Compositional engineering of multicomponent garnet scintillators: Towards an ultra-accelerated scintillation response. *Mater. Adv.* **2022**, *3*, 6842–6852. [CrossRef]
10. Zhu, D.; Nikl, M.; Chewpraditkul, W.; Li, J. Development and prospects of garnet ceramic scintillators: A review. *J. Adv.Ceram.* **2022**, *11*, 1825–1848. [CrossRef]
11. Korzhik, M.; Retivov, V.; Dosovitskiy, G.; Dubov, V.; Kamenskikh, I.; Karpuk, P.; Komendo, I.; Kuznetsova, D.; Smyslova, V.; Mechinsky, V.; et al. First Observation of the Scintillation Cascade in Tb^{3+}-Doped Quaternary Garnet Ceramics. *Phys. Status Solidi R* **2023**, *17*, 2200368. [CrossRef]
12. Karpyuk, P.; Korzhik, M.; Fedorov, A.; Kamenskikh, I.; Komendo, I.; Kuznetsova, D.; Leksina, E.; Mechinsky, V.; Pustovarov, V.; Smyslova, V.; et al. The Saturation of the Response to an Electron Beam of Ce- and Tb-Doped GYAGG Phosphors for Indirect β-Voltaics. *Appl. Sci.* **2023**, *13*, 3323. [CrossRef]
13. Dubov, V.; Gogoleva, M.; Saifutyarov, R.; Kucherov, O.; Korzhik, M.; Kuznetsova, D.; Komendo, I.; Sokolov, P. Micro-Nonuniformity of the Luminescence Parameters in Compositionally Disordered GYAGG:Ce Ceramics. *Photonics* **2023**, *10*, 54. [CrossRef]
14. Retivov, V.; Dubov, V.; Kuznetsova, D.; Ismagulov, A.; Korzhik, M. Gd^{3+} content optimization for mastering high light yield and fast Gd$_x$Al$_2$Ga$_3$O$_{12}$:Ce^{3+} scintillation ceramics. *J. Rare Earths* **2023**, in press. [CrossRef]
15. Kuznetsova, D.; Dubov, V.; Bondarev, A.; Dosovitskiy, G.; Mechinsky, V.; Retivov, V.; Kucherov, O.; Saifutyarov, R.; Korzhik, M. Tailoring of the Gd–Y–Lu ratio in quintuple (Gd, Lu, Y)$_3$Al$_2$Ga$_3$O$_{12}$:Ce ceramics for better scintillation properties. *J. Appl. Phys.* **2022**, *132*, 203104. [CrossRef]
16. Dantelle, G.; Boulon, G.; Guyot, Y.; Testemale, D.; Guzik, M.; Kurosawa, S.; Kamada, K.; Yoshikawa, A. Research on Efficient Fast Scintillators: Evidence and X-Ray Absorption Near Edge Spectroscopy Characterization of Ce^{4+} in Ce^{3+}, Mg^{2+}-Co-Doped Gd$_3$Al$_2$Ga$_3$O$_{12}$ Garnet Crystal. *Phys. Status Solidi B* **2020**, *257*, 1900510. [CrossRef]
17. Korzhik, M.; Alenkov, V.; Buzanov, O.; Dosovitskiy, G.; Fedorov, A.; Kozlov, D.; Mechinsky, V.; Nargelas, S.; Tamulaitis, G.; Vaitkevicius, A. Engineering of a new single-crystal multi-ionic fast and high-light-yield scintillation material (Gd$_{0.5}$–Y$_{0.5}$)$_3$Al$_2$Ga$_3$O$_{12}$:Ce,Mg. *CrystEngComm* **2020**, *22*, 2502–2506. [CrossRef]
18. Spassky, D.; Kozlova, N.; Zabelina, E.; Kasimova, V.; Krutyak, N.; Ukhanova, A.; Morozov, V.A.; Morozov, A.V.; Buzanov, O.; Chernenko, K.; et al. Influence of the Sc cation substituent on the structural properties and energy transfer processes in GAGG:Ce crystals. *CrystEngComm* **2020**, *22*, 2621–2631. [CrossRef]
19. Shi, Y.; Shichalin, O.; Xiong, Y.; Kosyanov, D.; Wu, T.; Zhang, Q.; Wang, L.; Zhou, Z.; Wang, H.; Fang, J.; et al. Ce^{3+} doped Lu$_3$Al$_5$O$_{12}$ ceramics prepared by spark plasma sintering technology using micrometre powders: Microstructure, luminescence, and scintillation properties. *J. Eur. Ceram. Soc.* **2022**, *42*, 6663–6670. [CrossRef]
20. Kuznetsov, S.V.; Sedov, V.S.; Martyanov, A.K.; Batygov, S.C.; Vakalov, D.S.; Boldyrev, K.N.; Tiazhelov, I.A.; Popovich, A.F.; Pasternak, D.G.; Bland, H.; et al. Cerium-doped gadolinium-scandium-aluminum garnet powders: Synthesis and use in X-ray luminescent diamond composites. *Ceram. Int.* **2022**, *48*, 12962–12970. [CrossRef]
21. Fedorov, A.; Komendo, I.; Amelina, A.; Gordienko, E.; Gurinovich, V.; Guzov, V.; Dosovitskiy, G.; Kozhemyakin, V.; Kozlov, D.; Lopatik, A.; et al. GYAGG/^{6}LiF composite scintillation screen for neutron detection. *Nucl. Eng. Technol.* **2022**, *54*, 1024–1029. [CrossRef]

22. Rudzik, T.J.; Seeley, Z.M.; Drobshoff, A.D.; Cherepy, N.J.; Wang, Y.; Onorato, S.P.; Squillante, M.R.; Payne, S.A. Additively manufactured transparent ceramic thin disk gain medium *Opt. Mater. Express.* **2022**, *12*, 3648–3657. [CrossRef]
23. Osborne, R.A.; Wineger, T.J.; Yee, T.D.; Cherepy, N.J.; Seeley, Z.M.; Gaume, R.; Dubinskii, M.; Payne, S.A. Fabrication of engineered dopant profiles in Er/Lu:YAG transparent laser ceramics via additive manufacturing. *Opt. Mater. Express.* **2023**, *13*, 526–537. [CrossRef]
24. Ermakova, L.V.; Dubov, V.V.; Saifutyarov, R.R.; Kuznetsova, D.E.; Malozovskaya, M.S.; Karpyuk, P.V.; Dosovitskiy, G.A.; Sokolov, P.S. Influence of Luminescent Properties of Powders on the Fabrication of Scintillation Ceramics by Stereolithography 3D Printing. *Ceramics* **2023**, *6*, 43–57. [CrossRef]
25. Fedorov, A.A.; Dubov, V.V.; Ermakova, L.V.; Bondarev, A.G.; Karpyuk, P.V.; Korzhik, M.V.; Kuznetsova, D.E.; Mechinsky, V.A.; Smyslova, V.G.; Dosovitskiy, G.A.; et al. $Gd_3Al_2Ga_3O_{12}$:Ce Scintillation Ceramic Elements for Measuring Ionizing Radiation in Gases and Liquids. *Instrum. Exp. Tech.* **2023**, *66*, 234–238. [CrossRef]
26. Karipbayev, Z.T.; Lisitsyn, V.M.; Mussakhanov, D.A.; Alpyssova, G.K.; Popov, A.I.; Polisadova, E.F.; Elsts, E.; Akilbekov, A.T.; Kukenova, A.B.; Kemere, M.; et al. Time-resolved luminescence of YAG:Ce and YAGG:Ce ceramics prepared by electron beam assisted synthesis. *Nucl. Instrum. Meth. B* **2020**, *479*, 222–228. [CrossRef]
27. Abd, H.R.; Hassan, Z.; Ahmed, N.M.; Omar, A.F.; Thahab, S.M.; Lau, K.S. Rapid synthesis of Ce^{3+}:YAG via CO_2 laser irradiation combustion method: Influence of Ce doping and thickness of phosphor ceramic on the performance of a white LED device. *J. Solid State Chem.* **2021**, *294*, 121866. [CrossRef]
28. Loiko, P.; Basyrova, L.; Maksimov, R.; Shitov, V.; Baranov, M.; Starecki, F.; Mateos, X.; Camy, P. Comparative study of $Ho:Y_2O_3$ and $Ho:Y_3Al_5O_{12}$ transparent ceramics produced from laser-ablated nanoparticles. *J. Lumin.* **2021**, *240*, 118460. [CrossRef]
29. Timoshenko, A.D.; Matvienko, O.O.; Doroshenko, A.G.; Parkhomenko, S.V.; Voronova, I.O.; Kryzhanovska, O.S.; Safronova, N.A.; Vovk, O.O.; Tolmachev, A.V.; Baumer, V.N.; et al. Highly-doped $YAG:Sm^{3+}$ transparent ceramics: Effect of Sm^{3+} ions concentration. *Ceram. Int.* **2023**, *49*, 7524–7533. [CrossRef]
30. Wagner, A.; Ratzker, B.; Kalabukhov, S.; Kolusheva, S.; Sokol, M.; Frage, N. Highly-doped Nd:YAG ceramics fabricated by conventional and high pressure SPS. *Ceram. Int.* **2019**, *45*, 12279–12284. [CrossRef]
31. Derdzyan, M.V.; Hovhannesyan, K.L.; Santos, S.N.C.; Dujardin, C.; Petrosyan, A.G. Influence of Air Annealing on Optical and Scintillation Properties of YAG:Pr,Ca. *Phys. Status Solidi A* **2022**, *220*, 2200571. [CrossRef]
32. Kuznetsova, D.E.; Volkov, P.A.; Dosovitskiy, G.A.; Mikhlin, A.L.; Bogatov, K.B.; Retivov, V.M.; Dosovitskiy, A.E. Influence of alkali metal impurities on properties of yttrium aluminum garnet doped with cerium. *Russ. Chem. Bull.* **2016**, *65*, 1734–1738. [CrossRef]
33. Karpyuk, P.; Shurkina, A.; Kuznetsova, D.; Smyslova, V.; Dubov, V.; Dosovitskiy, G.; Korzhik, M.; Retivov, V.; Bondarev, A. Effect of Sintering Additives on the Sintering and Spectral-Luminescent Characteristics of Quaternary GYAGG:Ce Scintillation Ceramics. *J. Electron. Mater.* **2022**, *51*, 6481–6491. [CrossRef]
34. Vorona, I.O.; Yavetskiy, R.P.; Parkhomenko, S.V.; Doroshenko, A.G.; Kryzhanovska, O.S.; Safronova, N.A.; Timoshenko, A.D.; Balabanov, A.E.; Tolmachev, A.V.; Baumer, V.N. Effect of complex $Si^{4+}+Mg^{2+}$ additive on sintering and properties of undoped YAG ceramics. *J. Eur. Ceram. Soc.* **2022**, *42*, 6104–6109. [CrossRef]
35. Setlur, A.A.; Heward, W.J.; Hannah, M.E.; Happek, U. Incorporation of $Si^{4+}–N^{3-}$ into Ce^{3+}-Doped Garnets for Warm White LED Phosphors. *Chem. Mater.* **2008**, *20*, 6277–6283. [CrossRef]
36. Dosovitskii, G.A.; Bogatov, K.B.; Volkov, P.A.; Mikhlin, A.L.; Dosovitskii, A.E. Effect of adding boron on morphological and functional properties of aluminum-yttrium garnet activated with europium. *Refract. Ind. Ceram.* **2013**, *54*, 69–73. [CrossRef]
37. Yan, Y.; Meshbah, A.; Kheouz, L.; Bouillet, C.; Lorentz, C.; Blanchard, N.; Berends, A.C.; van der Haar, M.A.; Lerouge, F.; Krames, M.R.; et al. Ultra-Small YPO_4-YAG:Ce Composite Nanophosphors with a Photoluminescence Quantum Yield Exceeding 50%. *Small* **2023**, *19*, 2208055. [CrossRef]
38. Pirrung, F.O.H.; Noordam, A.; Harbers, P.J.; Loen, E.M.; Munneke, A.E. Phosphoric Acid Esters and Their Use as Wetting and Dispersing Agent. US Patent 7595416B2, 28 February 2005. Available online: https://patents.google.com/patent/US7595416B2/ (accessed on 5 July 2023).
39. Gobelt, B.; Nagelsdiek, R.; Omeis, J.; Piestert, F.; Pritschins, W.; Meznaric, N.; Schroder, D.; Tiegs, W. Dispersing Additives Based on Phosphoric Acid Ester Derivatives. US Patent 9518146, 4 May 2012. Available online: https://patents.google.com/patent/US9518146B2 (accessed on 5 July 2023).
40. Gordienko, E.; Fedorov, A.; Radiuk, E.; Mechinsky, V.; Dosovitskiy, G.; Vashchenkova, E.; Kuznetsova, D.; Retivov, V.; Dosovitskiy, A.; Korjik, M.; et al. Synthesis of crystalline Ce-activated garnet phosphor powders and technique to characterize their scintillation light yield. *Opt. Mater.* **2018**, *78*, 312–318. [CrossRef]
41. Pardo, A.; Romero, J.; Ortiz, E. High-temperature behaviour of ammonium dihydrogen phosphate. *J. Phys. Conf. Ser.* **2017**, *935*, 012050. [CrossRef]
42. Su, C.H.; Chen, C.C.; Liaw, H.J.; Wang, S.C. The Assessment of Fire Suppression Capability for the Ammonium Dihydrogen Phosphate Dry Powder of Commercial Fire Extinguishers. *Procedia Eng.* **2014**, *84*, 485–490. [CrossRef]
43. Li, X.; Wang, H.; Guan, L.; Fu, Y.; Guo, Z.; Yuan, K.; Tie, L.; Yang, Z.; Teng, F. Influence of pH value on properties of $YPO_4:Tb^{3+}$ phosphor by co-precipitation method. *J. Rare Earths* **2015**, *33*, 346–349. [CrossRef]
44. Li, J.; Zhu, J.; Li, P.; Nan, C.; Zhang, Y. Roles of $(NH_4)_2HPO_4$ and $NH_4H_2PO_4$ in the phase transition and luminescence enhancement of $YPO_4:Eu$. *J. Phys. Chem. Solids* **2021**, *150*, 109821. [CrossRef]
45. Agrawal, D.; Hummel, F.A. The Systems Y_2O_3-P_2O_5 and Gd_2O_3-P_2O_5. *J. Electrochem. Soc.* **1980**, *127*, 1550. [CrossRef]

46. Szuszkiewicz, W.; Znamierowska, T. Y_2O_3-P_2O_5 Phase Diagram. *Pol. J. Chem.* **1989**, *63*, 381–391.
47. Grew, E.S.; Locock, A.J.; Mills, S.J.; Galuskina, I.O.; Galuskin, E.V.; Halenius, U. Nomenclature of the garnet supergroup. *Am. Miner.* **2013**, *88*, 785–811. [CrossRef]
48. Thompson, R.N. Is upper-mantle phosphorus contained in sodic garnet? *Earth Planet Sci. Lett.* **1975**, *26*, 417–424. [CrossRef]
49. Brunet, F.; Bonneau, V.; Irifune, T. Complete solid-solution between $Na_3Al_2(PO_4)_3$ and $Mg_3Al_2(SiO_4)_3$ garnets at high pressure. *Am. Mineral.* **2006**, *91*, 211–215. [CrossRef]

Disclaimer/Publisher's Note: The statements, opinions and data contained in all publications are solely those of the individual author(s) and contributor(s) and not of MDPI and/or the editor(s). MDPI and/or the editor(s) disclaim responsibility for any injury to people or property resulting from any ideas, methods, instructions or products referred to in the content.

MDPI
St. Alban-Anlage 66
4052 Basel
Switzerland
www.mdpi.com

Ceramics Editorial Office
E-mail: ceramics@mdpi.com
www.mdpi.com/journal/ceramics

Disclaimer/Publisher's Note: The statements, opinions and data contained in all publications are solely those of the individual author(s) and contributor(s) and not of MDPI and/or the editor(s). MDPI and/or the editor(s) disclaim responsibility for any injury to people or property resulting from any ideas, methods, instructions or products referred to in the content.

www.ingramcontent.com/pod-product-compliance
Lightning Source LLC
LaVergne TN
LVHW070630100526
838202LV00012B/774